GAIN

The MIT Press Essential Knowledge Series

A complete list of books in this series can be found online at
https://mitpress.mit.edu/books/series/mit-press-essential-knowledge-series.

GAIN OF FUNCTION

NICHOLAS G. EVANS

The MIT Press | Cambridge, Massachusetts | London, England

© 2025 Massachusetts Institute of Technology

All rights reserved. No part of this book may be used to train artificial intelligence systems or reproduced in any form by any electronic or mechanical means (including photocopying, recording, or information storage and retrieval) without permission in writing from the publisher.

The MIT Press would like to thank the anonymous peer reviewers who provided comments on drafts of this book. The generous work of academic experts is essential for establishing the authority and quality of our publications. We acknowledge with gratitude the contributions of these otherwise uncredited readers.

This book was set in Chaparral Pro by New Best-set Typesetters Ltd. Printed and bound in the United States of America.

Library of Congress Cataloging-in-Publication Data

Names: Evans, Nicholas G., 1985- author.
Title: Gain of function / Nicholas G. Evans.
Description: Cambridge, Massachusetts : The MIT Press, [2025] | Series: The MIT Press essential knowledge series | Includes bibliographical references and index.
Identifiers: LCCN 2024019111 (print) | LCCN 2024019112 (ebook) | ISBN 9780262551359 (paperback) | ISBN 9780262381727 (epub) | ISBN 9780262381734 (pdf)
Subjects: MESH: Viruses—pathogenicity | Biomedical Research—ethics | Gain of Function Mutation | Virology—ethics | Dual Use Research—ethics | Public Policy
Classification: LCC QR360 (print) | LCC QR360 (ebook) | NLM QW 160 | DDC 616.9/101—dc23/eng/20241023
LC record available at https://lccn.loc.gov/2024019111
LC ebook record available at https://lccn.loc.gov/2024019112

10 9 8 7 6 5 4 3 2 1

In memoriam, Dr. Jo L. Husbands-Rosenberg
(1948–2022)

CONTENTS

Series Foreword ix

1 What Is Gain of Function? 1
2 Dual-Use Research in the Life Sciences 23
3 The Science of Gain of Function 45
4 The Emergence of Gain of Function as a Policy Debate 77
5 Gain-of-Function Policy 91
6 Current Controversies 115
7 The Ethics of Gain of Function 143
8 The Future of Gain of Function 159

Glossary 187
Notes 189
Further Reading 203
Index 211

SERIES FOREWORD

The MIT Press Essential Knowledge series offers accessible, concise, beautifully produced pocket-size books on topics of current interest. Written by leading thinkers, the books in this series deliver expert overviews of subjects that range from the cultural and the historical to the scientific and the technical.

In today's era of instant information gratification, we have ready access to opinions, rationalizations, and superficial descriptions. Much harder to come by is the foundational knowledge that informs a principled understanding of the world. Essential Knowledge books fill that need. Synthesizing specialized subject matter for nonspecialists and engaging critical topics through fundamentals, each of these compact volumes offers readers a point of access to complex ideas.

1

WHAT IS GAIN OF FUNCTION?

Gain of Function and Dual Use

This book was written during the coronavirus disease 2019 (COVID-19) pandemic: declared over by world leaders, but still burning in hospitals, congregate settings like nursing homes and prisons, our schools and workplaces, and our homes. When I was first asked to write this book, I had the sense that now was not the time for a book about what I thought was a niche area of interest in science policy occupied by a couple dozen—at most—scholars and activists, including myself. Surely there were more important things to discuss, research, and push for in terms of government policy.

But the issue in question has exploded onto our social media, in our news outlets, and even into the halls of the US Capitol Building. In 2021, Senator Rand Paul accused

Anthony Fauci of lying to the American people and concealing evidence that the US government allegedly funded "gain-of-function" (GOF) research in Wuhan, China—a purported origin story of COVID-19. Over the course of the next two years, the senator would threaten Fauci with imprisonment, the National Institutes of Health (NIH) with defunding, and the life sciences with being upended as a handmaiden to a pandemic.

Beyond the halls of Congress and its more colorful inhabitants, the relationship between this handful of scientific studies and the pandemic continues to be described in a detail that I can only call "cinematic" for its color along with the way the framing of the pandemic's origins serves as a lens that focuses the eye on some details and occludes others. An entire distributed investigative ecosystem, whose most prominent group is known as DRASTIC, has sprung up to determine exactly which study is the smoking gun for COVID-19. My library shelves groan with the increasing number of books that purport to show, beyond a reasonable doubt, that COVID-19 came from a lab—and more, that it was *built* in a lab. We don't seem any closer to the truth, but that's never stopped business, and business is good for pandemic entrepreneurs.

As someone who finds that the preponderance of evidence heads in the direction of a natural origin to the pandemic—or as natural as it can be given the rapid encroachment on the habitats of animals that carry zoonotic pathogens, whose ranges are changing in response to anthropogenic climate change—this all seemed ridiculous on its face. And in the decade I have interacted with NIH staff and officials at a variety of levels, often as an interlocutor about this very issue, I've never known them to lie in such a manner. The NIH is labyrinthine, and bureaucratic to a fault, but it's generally not deceptive—they, like many federal employees, know how to work around a Freedom of Information Act request, but they don't outright lie. And life scientists, for all the shapes and sizes they come in, are typically not the kind to organize a world-shaking conspiracy. So perhaps it *is* best we lay out what GOF research is and is not, and what it means for science, governance, politics, and our values.

But we're getting ahead of ourselves.

The life sciences—and by that I mean scientific fields focused on the natural organic world, from virology and immunology, to zoology and ecology, to epidemiology and medical research—are generally thought to be a huge boon to humanity. The Salk vaccine eliminated polio in most of the world and, despite several setbacks, we are tantalizing close to *eradication* of the disease. This builds on the success of smallpox eradication, the only human pathogen ever to be extinguished by political will and scientific ingenuity. The discovery of recombinant DNA in the 1970s lead to the synthetic production of insulin, rendering diabetes an imminently manageable disease, only lethal because of the failures of national and global health systems. Antibiotics have made surgeries lifesaving instead of life-threatening, while they remain effective. And broad public health interventions such as fluoridation and sanitation have improved the health of billions of people worldwide, when we have the will to use them.

Within the life sciences there are only a small number of technologies and discoveries that have the opposite capacity—to harm rather than heal. But in between are a good number of life sciences discoveries that can help *or* harm humanity, and some of these can do so on a grand scale. These discoveries are often referred to as "dual-use" research, where the "dual" is not the number of uses but instead their valences: the good and bad.

GOF research is one example of dual-use research. The basic idea behind GOF research is to take a virus or other pathogen, and through a variety of scientific means enhance how deadly it is, how fast it spreads, or which kind of species it infects—its virulence, transmissibility, or host range. This kind of research is the subject of controversy because some claim that it is necessary to defend against pandemic diseases like COVID-19, while others—including, for the last decade, myself—argue that in key cases it poses risks to humanity that exceed its benefits and, even if necessary, should only be conducted if no other experiment answers the same critical public health questions.

This book is about GOF research: what it is and is not, how it works, why it emerged as an issue of policy on a national and international level, and the ethical and policy considerations that we can use to think through its funding, conduct, and publication. Why is a professor of philosophy writing such a book? Because for my entire career, from my graduate studies (including a dissertation on dual-use research) through to my latest publications, I have thought and written on GOF research. I've participated in some of the thornier debates in the United States and abroad on governing this kind of research. I know the players, literature, and details of this controversy. I have not "discovered" GOF research in the last couple of years on the back of varying controversies about the origins of COVID-19. And when the debates of today are settled or abandoned for the issues of tomorrow, I will still be thinking about GOF, what we do about it, and how it functions as a lodestone for larger debates about science and society.

The Paradigm Cases

Let's start with what philosophers like to call "paradigm cases." These cases are critical to any serious thinking about scientific, policy, and ethical issues because they frame our ideas as a starting point from which to think about everything else.[1] In the case of GOF, there are two paradigm cases, and they happened at the same time.

Ron Fouchier

On November 23, 2011, an article appeared in the news section of *Science* magazine. Better known as a peer-reviewed scientific journal, *Science* is also a news outlet of a certain kind: the media arm of the American Association for the Advancement of Science, one of the oldest trade organizations in the United States.[2] This article carried the headline "Scientists Brace for Media Storm around Controversial Flu Studies." The first lines of the article capture the mood of the debate that would become GOF: "ROTTERDAM, THE NETHERLANDS—Locked up in the bowels of the medical faculty building here and accessible to only a handful of scientists lies a man-made flu virus that could change world history if it were ever set free."[3] The virus in question is known as highly pathogenic avian influenza H5N1 (you will sometimes read this as "HPAI H5N1" as there are *low* pathogenicity versions of the virus; I'm going to drop the HPAI prefix for ease of reading), sometimes called "bird flu." Almost all influenza A viruses are avian in origin, making most flus "bird flus" in one sense or another. What makes H5N1 special is that right now it undergoes transmission only between birds. Humans can *get* H5N1, but they can't *spread it*. Of those who do get it, however, more than half have died.

The study the article described was led by Ron Fouchier, a luminary in influenza biology. I call it a study because it wasn't "an experiment," singular, but rather a series of seven experiments the research team undertook to create a form of H5N1 that would transmit in mammals. Not humans, mind you. Instead, the researchers showed that their new form of H5N1 could transmit between ferrets, the stand-in for humans in influenza research (see figure 1). One of the seven experiments they conducted is referred to as "serial passaging," where a virus is inserted into an animal, allowed to replicate, and then extracted and implanted into the next. In the case of ferrets, this was done through their noses. After enough cycles of this implant-extract process, the virus had replicated and

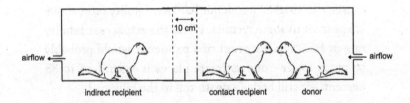

Figure 1 One kind of serial passaging, involving transmission. In some experiments, the "donor" and "recipient" of the virus are separated in different enclosures; in others, they are housed together. One animal (here, a ferret) has the virus, and the other does not. This is repeated a number of times in a particular experiment. Image courtesy of Mathilde Richard et al., "SARS-CoV-2 Is Transmitted via Contact and via the Air between Ferrets," *Nature Communications* 11, no. 1 (July 8, 2020): 3496, https://doi.org/10.1038/s41467-020-17367-2.

evolved in the ferrets until it could spread from ferret to ferret just through their breathing and sneezes.[4]

Yes, ferrets sneeze, and it is quite cute when it doesn't involve what Fouchier alleged in the same *Science* news article he had done, which sounded like the start of a disaster movie: "In a 17th floor office in the same building, virologist Ron Fouchier of Erasmus Medical Center calmly explains why his team created what he says is 'probably one of the most dangerous viruses you can make.'" The controversy, then, was that scientists took a strain of influenza that to date has killed 60 percent of the people who have contracted it and gave it the ability to *potentially* spread between humans. For reference, COVID-19 has a "case fatality rate" of approximately 1 percent, which means it kills 1 in every 100 people it infects. The flu virus that caused the 1918 flu pandemic had case fatality rates above 2.5 percent in some records. While the actual case fatality rate of H5N1 in the event of a pandemic would probably be much lower—one estimate places it around 14 to 33 percent—it still bodes a death toll in the billions.[5]

Of course, nothing is ever that simple. One of the things the researchers found was that even though the host range—remember, the kinds of animals the virus infects and spreads between—had increased, the virulence seemed to decrease. No ferrets died from the virus in these experiments. Of course, there were only 10 ferrets in the experiment; if the death rate were 60 percent, we would

COVID-19 has a
"case fatality rate" of
approximately 1 percent,
which means it kills
1 in every 100 people
it infects.

expect 6 to die, but because there were only 10 ferrets, if the eleventh had died that would have made the death rate in the experiment 9 percent, which is still alarming. And there is a chance that every ferret in the experiment simply rolled the right dice—less than a 1 percent chance they *all* survived a 60 percent chance of death, but still in the realm of possibility.

Yoshihiro Kawaoka

At the same time, another experiment was conducted by researchers at the University of Wisconsin at Madison, under the leadership of Yoshihiro Kawaoka, another giant of the world of influenza virology. That team didn't just do serial passaging. Rather, it engaged in a "reassortment" experiment: the team pieced together a virus from the backbone of H5N1 and the parts of the 2009 pandemic H1N1 flu virus (you may remember it as "swine flu") that allow influenza to enter and exit cells. These parts are called hemagglutinin and neuraminidase, and are the "H" and "N" in names like H5N1, H1N1, H7N9, and so on. The researchers then implanted the virus into ferrets and allowed it to sort itself out into a virus that was stable and, again, spread from ferret to ferret.[6]

The two studies *do* capture some crucial aspects of the spread of infectious diseases that can make them more dangerous. The first is the capacity for viruses to evolve as they jump from person to person—something we are all deeply familiar with given the spread of names for COVID-19 variants that continues to this day. The second is the capacity for viruses to trade biological material when present in the same host. Flu viruses are particularly infamous for this, such as when a virus from a bird and a virus from a human end up in a pig and combine to form something new.

The concern at the time, however, was that even if this were good science and valuable to our understanding of pandemic influenza, creating a virus like this posed a *security* risk. Government officials, commentators, and other scientists worried that while Fouchier's and Kawaoka's teams were well known and well intentioned, not everyone is. Putting this kind of research out into the world posed what some call an "information hazard" because it provided a road map for a bioterrorist to create a weapon.[7]

Is Everything GOF in Biology?

GOF experiments, so far, sound pretty generic. The life sciences are packed with experiments involving sticking a microorganism into an animal or messing with its genes and seeing what happens. Sometimes the organism thrives, and in new and exciting (or scary) ways. Sometimes it dies, its genetic code selected against by harsh evolutionary pressures. Does that make everything GOF?

Even if this were good science and valuable to our understanding of pandemic influenza, creating a virus like this posed a *security* risk.

This isn't an idle question. In the wake of our paradigm cases in 2011, and in future episodes I'll cover later, some scientists resisted the idea of doing anything about "GOF research" because of the breadth of the term. In 2014, Kanta Subbarao—who now runs the WHO Collaborating Center for Reference and Research on Influenza in Melbourne, Australia—spoke at the US National Academies of Sciences, explaining that

> routine virological methods involve experiments that aim to produce a gain of a desired function, such as higher yields for vaccine strains, but often also lead to loss of function, such as loss of the ability for a virus to replicate well, as a consequence. In other words, any selection process involving an alteration of genotypes and their resulting phenotypes is considered a type of Gain-of-Function (GoF) research, even if the U.S. policy is intended to apply to only a small subset of such work.[8]

The upshot, then, is that any attempt to control science that might look like Fouchier's or Kawaoka's would inevitably crush, or at least risk crushing, mainstream biological research that benefits us all. The implication is that virologists have been talking about GOF for decades, and that the US government moving in on the term is misguided and worrying.

A natural response to this, then, is to head to the library. The National Library of Medicine—arguably one the most authoritative collections of life sciences literature in the world—has about 18,000 documents that mention "gain of function" between 1982 and 2023. It also has over a million entries in virology, dating back to 1943. But GOF *papers*, it turns out, have nothing to do with viruses—only 541 virology papers mention the term.[9] The rest of those papers are about, for example, *cancer* genes, not virus genes. And I've yet to see, though can't rule out the possibility of, oncologists participating in the GOF debate.

Drill down into these results further, and things get weirder. About 197 results are published in 2011 or earlier—before the Fouchier and Kawaoka papers became common knowledge in November of that year. This is out of the 396,179 papers found in a search for virology and viruses published between 1982 and 2011: 0.05 percent, or 1 in 2,000 papers. Hardly commonplace. A similar search for influenza and "gain of function" yields 4 out of 20,866 results, or 0.02 percent. The point is not that the term isn't used but rather that it's implausible to imagine that—as appeared in an op-ed written for the medical news outlet *STAT* in 2021 by a former president of the American Society of Microbiology and the current CEO of that organization—scientists in the 1970s and 1980s studying the insertion of pig genes into *E.coli* to synthesize insulin would have been doing "gain-of-function research," either in the eyes of the US government or in their own language.[10] It is likewise simply not true in the "call for a rational discourse" made by some of the world's leading life scientists that "gain-of-function approaches incorporate a large proportion of all research because they are a powerful genetic tool in the laboratory."[11]

Even the most expansive view still gives a minority position on GOF in *biology*. After conducting a search of the National Library of Medicine for journal articles published after 2000 that involved animals or animal diseases—or any virus, infection, or pathogen—researchers at Georgetown found that approximately 5 percent of all studies involved a "gain of function" or "loss of function." Yet what they counted *as* a gain (or loss) was any study that *resulted* (intended or not) in the enhancement or diminishment of a pathogen's replication, survival, symptoms, transmission, ability to evade the immune system, or resistance to drugs including vaccines. This means that, to use one of their examples, injecting mice with the bacterium *Pseudomonas aeruginosa*, which normally doesn't like mice, to see how it interacts with the mouse immune system is GOF research. From that 5 percent, around half involves a *gain* of a function, meaning that by their count, something like 3,500 papers, 2.5 percent of their sample, in the last 22 years have involved a gain of a function.[12]

This can tell us a few things about GOF research. First, we are typically talking about *studies* and not just single experiments involving a simple addition of genes in life sciences research leading to some consequence. Some GOF studies may not identify the genes that cause the change in behavior in a virus, such as a serial passaging experiment that simply passes a virus through animal hosts until it can reliably transmit between them. Sometimes GOF studies suggest a gene, but other experiments contradict them. The kind of study, however, remains the same: take a virus, and enhance its properties in specific ways that involve transmissibility, virulence, or host range.

The second property of a GOF study is the reason it is conducted. Some experiments use GOF genes to ask questions about the biology of cancers, or viruses, or other things. But *GOF studies* seek to produce a modified virus that we expect to pose serious risks to human health. That is their purpose, and while they may have other purposes or results, the creation of the modified virus is intentional. In the paradigm cases, the modified viruses were novel in that they were not believed to have yet emerged during influenza's near-constant genetic reshuffling in nature. This may actually be a weakness of these studies because it's not clear these viruses could actually arise in nature—in which case, studying them might be a scientific dead end.[13]

Third, while a lot of experiments seem on first blush to involve a "gain" of "function," and indeed it may be true that some scientists have talked about microbiological experiments leading to a gain of function, that phrase hardly seems to have been a *term of art* in virology until it became a policy term in 2011. I've no doubt that folks concerned about the definition have used the words "gain," "of," and "function" in sequence, or as a phrase, but it's hard to believe based on the publication record that they had a particularly well-defined idea of it. As a philosopher writing this book, I'll note that a lot of the hand-wringing about the definition seems to involve an overreach in the type of function gained, but doesn't acknowledge that in biology, the idea of "function" can itself mean multiple different things in the life sciences, at different times, to different people.

So not everything is GOF. The term is not used that much in virology, at least in its literature; at most, 1 in 40 papers involves some gain of some function, and only around 15 percent of those are involved in vaccine research, one of the key issues for scientists who support the research. Sometimes experiments involve *a* GOF, but that experiment isn't a GOF *study*. But what is most significant here is that what a virologist means by GOF may be quite different from what a policy about GOF research means by it—but in a way that is in practice easily understandable.

That shouldn't surprise or even concern us. We are all intimately familiar with these differences in language. No one who complained about unaffordable medical care was confused by the name of the Affordable Care Act, much less confused in cases where that act did not make care more affordable for some. People can tell the difference.

It's exceedingly common in policy circles that we must stipulate definitions, and the English language provides us with only so many combinations to choose from that make sense. While naming something in policy comes with risks, we've got a few thousand years of history of governance to navigate that—for better and worse. There's no reason to believe virologists have a monopoly on the term "GOF," especially because cancer researchers would probably have something to say about *that* in turn.

Policy Terms of Art and Scientific Concepts

This concern about definitions, and the way we use language, drives to the core of the GOF debate and everything that comes next in this book. The GOF debate is fundamentally about policy. And while the biology behind these experiments is necessary, it is only one piece in a more complex story about how we regulate the life sciences. Most everyone agrees that we should, if we can, promote the good in biology while avoiding the bad. That's not the controversial bit; how we go about that, and what counts as the good and bad of biology, is where the wicket gets exceedingly sticky.

With this in mind, throughout the rest of the book when I refer to GOF, I will do so as a *policy term*. That is, GOF describes, for our purposes,

> A scientific study that is reasonably expected to
> create a pathogen (bacteria, fungus, etc.) with an
> enhanced virulence, transmissibility, or host range
> from its natural template to such an extent that it
> poses a large-scale threat to human, animal, or plant
> populations.

This definition is my own, but it loosely tracks the debate from 2011 onward, the contours of a risk-and-benefit assessment on GOF research conducted in 2015, and the ethical and public health stakes of the issue. Before going further, however, a few clarifications. By "natural template," I mean the wild strain of virus the research team begins with. So in the case of the Fouchier team's experiment, it was H5N1, and specifically a strain from Indonesia from 2005. These are the viruses that are then modified into something new and often more terrible than the original.

By "a large-scale threat to human, animal, or plant populations," I mean a lethal and widespread disease epidemic. Its scope will typically be broad but need not be a "pandemic," with all the controversies of that term, as disease outbreaks can have global impacts without physically spreading to all four corners. And not every serious epidemic is a WHO-defined "public health emergency of international concern," which historically have only been declared on human pathogens. We could imagine a GOF experiment whose results are used to wipe out the coffee bean, which would be economically and ecologically devastating, and also harmful to human health as a consequence of these effects.[14]

Why choose this definition, which deals with policy rather than science? It might seem that the scientific term would be more precise or track something closer to the truth than something written by a civil servant. But the previous section shows the lie of that. GOF in science describes genes and experiments, and those experiments can be incredibly diverse! It describes cancer research, virology research, and more besides—fields that may have little or nothing to do with each other. And in the increasingly complex world of genomics, it is unlikely there are many true "GOF genes," as the idea of one gene, one function is largely dying away as we discover more about genetics as well as the interaction between genes, the environment, and other biological machinery. The level at which we understand and describe the "function" of an organism is thus also changing, which will complicate the scientific landscape in turn.

The policy definition I gave above, on the other hand, is hardly permanent—as we'll see, the newest policy on dual-use research attempts to move away from GOF altogether by wrapping it into a larger category of research—but instead captures something we should all really care about. That is, a small number of experiments, done for specific reasons, that have the potential to help or harm the globe in particular ways. And it is small; most experiments involving genetic manipulation of viruses don't use especially dangerous viruses, don't create dangerous viruses, and don't *intend* to do either besides. This conception of GOF is narrow yet significant from the perspective of science, policy, and ethics.

In the next chapter, I'm going to take us back to the prehistory of GOF, before the term came to the fore in late 2011. This prehistory is of dual-use research in the life sciences and is an important framing device for later GOF discussions that eschew, and then later rediscover, the security aspects of the life sciences.

2

DUAL-USE RESEARCH IN THE LIFE SCIENCES

Dual-use research arises when one and the same piece of scientific research can be used to help and harm humanity. That's the definition, with little variation, for a small but steady literature on the regulation of life sciences that focus on research and technologies that potentially could become national as well as global security threats.[1] The "global" here is important, as is the idea of "harming humanity"; these are not technologies that could merely harm individuals but that could destroy nations too. And because the global is so crucial, in this chapter I will eschew the more complex definition adopted by the National Science Advisory Board for Biosecurity (NSABB) of "dual-use research of concern" (DURC), which is "research that, based on current understanding, can be reasonably anticipated to provide knowledge, products, or technologies that could be directly misapplied by others to pose a threat to public health and safety, agricultural crops and other plants, animals, the environment, or material."[2]

This is the prehistory of GOF, and the environment into which the controversy about Fouchier's and Kawaoka's papers would emerge. It is important to know this history for two reasons. The first is that the history of GOF is a history of the policy responses to the life sciences. The second is that with each new case of dangerous research, there is often a considerable amount of reinventing the wheel as researchers, activists, and policymakers "discover" an issue. But these issues are not discovered so much as new people come to an area and mistake what is novel for them as what is novel for the rest of the world.

To keep this short, I'll only deal with issues of dual use in the twenty-first century. There are arguably parallels in the life sciences back to the discovery of recombinant DNA in the 1970s, and the conferences at the Asilomar Conference Grounds in California that would ultimately lead to the first self-imposed regulations on genetic manipulation in the United States—though back then the concerns were largely about safety and legal liability, and not security. There are also parallels between the life sciences in the twenty-first century, and the experiences of physicists of the mid-twenty-first century and development of nuclear energy.[3] But these are much larger discussions than are necessary for our purposes here.

What Is "Dual Use" about Biology?

The term "dual use" is used, in discussions of science and security, in three related ways.[4] The first and oldest way this term is used is to describe technology that may be employed in military or civilian contexts. Things like GPS satellites, chipsets for supercomputers, and nuclear energy are paradigms of this kind of dual use. Note that these can be pieces of research, but you'll often hear these described as dual-use *technologies*, not dual-use *research*.

Dual-use *research* is the second kind of way "dual use" is used, and it is here that conversations about biology emerged in 2001. This kind of dual use is concerned with good and bad uses of research, and almost always with the "bad uses" being the creation of biological weapons or similar kinds of technologies of mass destruction. While the process from research to a functioning, mature weapon is long and arduous, most discussions of dual use in the life sciences are concerned with the research stages of biology.

The third and final kind is somewhat boutique, but emerges where the first and second collide. That sense of dual use concerns itself with *offensive* versus *defensive* uses of technology. A famous example of this in the early twentieth century was called Project Bacchus, which the US Defense Advanced Research Projects Agency (DARPA) funded to see if a bioweapon could be made with off-the-shelf components. It and its sister projects—Clear Vision, the reverse engineering of Soviet bioweapons technologies, and Jefferson, reproducing Soviet-made genetically modified anthrax—were touted as potential violations of the Biological and Toxin Weapons Convention, but were defended by the United States as justified for defensive reasons: figuring out how a weapon is made so as to defend against it.[5]

What unites all three is a concept of science and security that has its popular origins in the Manhattan Project: that we want science to help us, but some kinds of science help us best if *only we have access to them*, or if we at least have them *first*. This idea of scientific supremacy is surprisingly old. Similar discussions arose in revolutionary France, for example, around the development of new explosive chemicals for use in artillery and the accompanying concerns about the role of fortifications to defend against these new cannons.[6] Yet our story of dual use in the life sciences, canonically, begins not in the United States or in France but instead in a government lab in Australia.

Super Mousepox

Our first story is usually taken to be about mice, but is in fact a story about rabbits. A widely accepted tale in Australia goes that in 1859, Thomas Austin ordered 13 European wild rabbits to be shipped to his estate in the recently

This idea of scientific supremacy is surprisingly old.

settled colony of Victoria on Australia's south coast. The rabbits became a problem for Australia and Australians immediately. In 30 years, the rabbits colonized the entire country—faster, even, than Europeans—and by the 1940s there were 600 million rabbits living on the continent. After the failure of mechanical efforts such as fences and a 7-year bounty hunting campaign for rabbits in New South Wales, biological control became a potential cure for the rabbit plague. It would ultimately be the myxoma virus that would be deployed against the rabbits that had devastated the Australian ecosystem as a "grey blanket."[7]

The plan was not without its critics. The proposed release coincided with an unrelated outbreak of human encephalitis, and ambient public fears bled over into worry that humans might become sickened by the wild, intentional release of a poxvirus. Ian Clunies Ross, the chair of Australia's Commonwealth Scientific and Industrial Research Organisation (CSIRO), and scientists Macfarlane Burnet and Frank Fenner inoculated themselves with myxoma as a publicity stunt to demonstrate the safety of the virus in humans. Their good health was a sign that the viral campaign could go forward.

The first try seemed a remarkable success, killing 90 percent of the rabbits. The success was short-lived, however: by the end of the decade, rabbit populations began to climb as the surviving animals passed on genetic resistance to myxoma to their progeny. Fenner, better known for his work eradicating smallpox, immediately recognized the potential for the myxoma virus to select out a small population of resistant animals that could recover from the virus. Fortunately, a new virus would come along in the form of a rabbit hemorrhagic disease virus, which has evolved with its hosts to sustain relatively low rabbit populations. But even then, a new tool would ultimately be needed, and it was Ronald Jackson of CSIRO and Ian Ramshaw at the Australian National University—the latter, one of Fenner's graduate students—who would attempt to overcome the resistance rabbits had to myxoma. They would go one step further than nature and modify the virus to cause infertility in rabbits, depriving them of that crucial ability to develop resistance and then pass it on to their numerous progeny.

Myxoma is a relatively understudied virus, so Jackson and Ramshaw used the better-known *ectromelia variola*, also known as mousepox, as a model for their experiments. The pair saw initial success with mousepox, and their modified virus caused infected mice to remain infertile for between five and nine months after a single infection. Seeking a higher level of infectivity and the possibility of overcoming future genetic resistance to the modified virus, the team further modified *ectromelia* to express the interleukin-4 protein, a cytokine that plays a key role in mammalian immune systems. By encouraging the production of interleukin-4, the hope was that the mouse's immune system would actually assist the virus in the attack against its host's reproductive system. Their new virus, nicknamed *ectromelia variola* interleukin-4—or, prophetically, EVIL-4—encouraged the expression of interleukin-4 in mice as it attacked their fertility. Rather than render the mice infertile, though, it killed them. It killed, moreover, with horrifying efficiency: 100 percent of normal mice and vaccinated mice as well as 60 percent of mice already genetically resistant to *ectromelia* died with exposure to the new virus.[8]

Mousepox doesn't infect humans, so the new virus wasn't a direct threat to human health. Yet smallpox, a cousin of mousepox, does kill people. The "mousepox study" described a series of genetic changes that, if applied to smallpox, could potentially produce a virus with 100 percent lethality, and no cure or vaccine. Until 2014, when six vials of smallpox were discovered in a freezer in an old Food and Drug Administration building on the campus of the NIH in Bethesda, Maryland, it was thought that the virus only survived under lock and key at the Centers for Disease Control (CDC) in Atlanta, Georgia, and VECTOR, the Russian equivalent of the CDC in Sverdlovsk.[9] But the possibility of such an event was serious enough that even as they got around to submitting to a journal, Jackson and Ramshaw shared doubts with each other about what they'd done.

The mousepox study came about as concern over biological weapons turned into outright terror. In September 2001, a series of letters laced with anthrax were mailed to senators and media figures around the United States. Twenty-two people were infected; five died. Though an investigation into the strain and method of delivery was inconclusive as to the identity of the attacker, the incident highlighted the ease with which bioterror attacks could be pursued.

Jackson and Ramshaw were beset with accusations and interrogations. Should the study have been published? Should it have been conducted at all? Why, when the dangers of smallpox were so great—and the threat of bioterrorism so real—should scientists conduct dangerous research? And how should the possible fallout—in the form of more dangerous science—be managed?[10] While discussion about the biosecurity ramifications raged in the offices of policymakers and the security establishment, science trundled on. The pursuit to control plague rodent populations continued, and in 2004, CSIRO scientists announced that Jackson and Ramshaw's work could indeed be applied to myxoma, and used to kill genetically resistant rabbits.

Jackson and Ramshaw—with Fenner, among others—also published further work that demonstrated the mutation they introduced could overcome cidofovir, an important antiviral against smallpox, and thus if applied to smallpox could form the basis for a bioweapon against which there was no defense. To date, no bioterror plot has surfaced (at least in the unclassified world) involving genetically modified smallpox or any other poxvirus that infects humans. The legacy of the mousepox study, however, is our dual-use dilemma: when one and the same piece of scientific research may be used to benefit or harm humanity. How we promote the promise of the life sciences while deterring and intervening against maleficent actors—and those who ought to know better—has profound implications for the way scientists conduct and publish scientific research.

Poliovirus Synthesis

The sister study to the mousepox study concerns the synthesis of polio. In August 2002, Jeronimo Cello, Aniko Paul, and Eckard Wimmer published a paper in *Science* describing the chemical synthesis of poliovirus complementary DNA (cDNA), and its transcription and replication into infectious poliovirus. The scientific achievement was the demonstration of the chemical synthesis itself and the creation of a virus from its chemical components.[11] The dual-use property of the study was also just so: the researchers, funded by DARPA, had demonstrated that, in principle, any biological agent could be synthesized, including those used in biological weapons.[12]

Until 2002, biologists had been unable to successfully synthesize a biological organism in the absence of a natural template. Put another way, you needed an existing copy of a virus (or bacterium, fungus, etc.) to create more of that virus. Cello and colleagues opened the possibility that, in future, life sciences research could proceed without requiring the collection and maintenance of large collections of samples. So long as one stored sufficient information about an organism's genetics and molecular structure, one could simply re-create it in a lab. Synthesis is now more commonplace in the life sciences and the biotechnology industry, but 2002 was the first demonstrated instance that it could actually be done.

The reason the poliovirus study is a natural companion to the mousepox study is that it addressed what—for a few short months—was considered to be the limiting factor in the creation of a recombinant poxvirus for use against humans. Smallpox is an eradicated disease and thus there is no chance of finding samples indigenous to a population; there is currently no one, and hopefully will never again be anyone, who is sick with smallpox. The synthesis of poliovirus, observers inferred, meant that a clock was started, counting down to the inevitable synthesis of smallpox. When that happened, any motivated state or nonstate actor with sufficiently developed infrastructure and talent could produce one of the most feared viruses on the planet, if not modify it to be particularly deadly.

When the poliovirus study was reported in the *New York Times*, concern over the emergence of this new dual-use research served as partial motivation for reviewing the role of science and technology in creating risks of bioterrorism.[13] In response, the National Research Council was charged with producing a report ultimately titled *Biotechnology Research in an Age of Terrorism*, or the "Fink Report" after its chair, Gerald Fink.[14] The report noted that its charge in the Fink Report was an extension of a series of reports on science and security, beginning with the 1982 *Scientific Communication and National Security*. With the exception of the first, however, the dominant role of these reports was understanding how science could advance counterterrorism and counter biological attacks. The Fink Report, however, responded to the mousepox and poliovirus studies along with their status as potential national security threats in their own right. (This book is dedicated to Jo L. Husbands, one of the tireless staff of the National Academies of Science, Engineering, and Medicine who made the Fink Report happen. Rest in power, Jo.)

The Fink Report would ultimately make seven recommendations. Its first recommendation was to educate the scientific community on dual use. The purpose of this education, the Fink Report claimed, was to make the scientific community aware of its responsibilities to promote the beneficial uses of science while mitigating the risks that dual-use research would become a tool of malevolent state or nonstate actors. (The term "state or nonstate actors" is a reference to the possibility that individuals or small groups of terrorists could use dual-use research for their own ends, and features heavily in scholarship and policy work on dual use.)

The second recommendation states that an existing regulatory component of US laboratory research, the Institutional Biosafety Committee, could be augmented to cover the problem of dual use. In particular, the committee recommended considering seven experiments that each constituted a dual-use threat that was worth further review. Drawn from examples termed "contentious research" by Gerald Epstein, these experiments are now known as seven "experiments of concern" that policymakers refer to when considering dual use.[15] The experiments of concern are those that:

1. **Would demonstrate how to render a vaccine ineffective.** This would apply to both human and animal vaccines. Creation of a vaccine-resistant smallpox virus would fall into this class of experiments.

2. **Would confer resistance to therapeutically useful antibiotics or antiviral agents.** This would apply to

therapeutic agents that are used to control disease agents in humans, animals, or crops. Introduction of ciprofloxacin resistance in *Bacillus anthracis* would fall in this class.

3. **Would enhance the virulence of a pathogen or render a nonpathogen virulent**. This would apply to plant, animal, and human pathogens. Introduction of cereolysin toxin gene into *Bacillus anthracis* would fall into this class.

4. **Would increase transmissibility of a pathogen**. This would include enhancing transmission within or between species. Altering vector competence to enhance disease transmission would also fall into this class.

5. **Would alter the host range of a pathogen**. This would include making nonzoonotics into zoonotic agents. Altering the tropism of viruses would fit into this class.

6. **Would enable the evasion of diagnostic/detection modalities**. This could include microencapsulation to avoid antibody-based detection and/or the alteration of gene sequences to avoid detection by established molecular methods.

7. **Would enable the weaponization of a biological agent or toxin**. This would include the environmental

stabilization of pathogens. Synthesis of smallpox virus would fall into this class of experiments.[16]

The third recommendation situated publishers as the responsible party for dealing with dual-use research. While the second recommendation concerned preexperimental review, publishers were called on by the Fink Report to act autonomously to prevent the publication of dual-use research that seriously threatened national security. In doing so, though, the committee affirmed National Security Decision Directive 189, written during the administration of Ronald Reagan, that all "fundamental" scientific research be published without restriction.

The fourth and fifth recommendations concerned the NSABB. In its fourth recommendation, the committee argued for the creation of a "National Science Advisory Board for *Biodefense*," with a structure that would mirror the later NSABB. In the fifth, the committee called for a review of physical containment, and urged that personnel training be conducted periodically by the NSABB. The implication was that, rather than interfere with publication, the central concern be with preventing access to or accidental release of dangerous biological agents.

The sixth recommendation concerned law enforcement. The committee advised that domestic and federal law enforcement as well as national security communities "develop new channels of sustained communication with the life science."

Instead of selecting out the life sciences as worthy of greater scrutiny by law enforcement, the life sciences were cast as an ally in the fight against bioterrorism. This undoubtedly emerged from tensions between scientists and law enforcement in the aftermath of the 2001 anthrax attacks, in which law enforcement made a number of unsubstantiated accusations of life scientists for their participation in bioterrorism, most famously against Stephen Hatfill.[17]

The seventh and final recommendation concerned harmonizing the international oversight of biosecurity. The committee noted that "any serious attempt to reduce the risks associated with biotechnology must ultimately be international [in] scope, because the technologies that could be misused are available and being developed throughout the globe." This harmonization included educational, law and policy design, and scientific norms around the review and dissemination of dual-use research. The committee further suggested that international bodies including the World Health Organization (WHO) and United Nations Educational, Scientific and Cultural Organization along with scientific organizations such as the International Council for Science, InterAcademy Panel on International Issues, and InterAcademy Council (now jointly the InterAcademy Partnership) engage in education and policy efforts. These organizations—the last, recently created when the report was issued—represented scientists and scientific academies from more than 100 nations.

The Resurrection of the 1918 Flu

The final story in our brief prehistory is that of the resurrection of the 1918 H1N1 pandemic influenza strain sometimes (and erroneously) called the "Spanish flu." The misnomer came about because Spanish newspapers were the first—and for some time, the only—press outlets reporting on the flu due to media censorship in the First World War, leading to the misunderstanding that Spain was the epicenter of the outbreak. The historical record shows the first case of the flu appeared in Kansas. This pandemic is arguably the most famous prior to COVID-19, killing some 50 to 100 million over a period of approximately 18 months.[18]

In 2005, researchers announced that they had pieced back together the 1918 pandemic flu strain. The work was a combination of efforts from around the United States: a team at the US Army Medical Research Institute for Infectious Diseases had conceived the project, a team at Mount Sinai Hospital had helped design the blueprint for piecing the virus back together, and the CDC was the site at which the final work was ultimately done. A later experiment would see that sequence turned into a functioning virus and tested on primates.[19]

The work was controversial for similar reasons to the 2011 GOF studies. On the one hand, the scientists in question and their supporters argued that sequencing the 1918 flu was a key step to understanding what makes pandemic influenza so deadly, identifying eight genes that contributed to the 1918 strain's unique trajectory. The CDC makes it clear in its description of the experiment that "learning from the past" is essential to prevent future pandemics.[20]

But critics of the study saw a blueprint and even method for piecing together a deadly virus that could kill tens or even hundreds of millions of people. Moreover, they criticized proponents for being disingenuous about the benefits of the research. They may have had a point: in its "learning from the past" section, the CDC struggles to articulate what the 1918 flu sequencing did to advance public health plans for pandemic influenza—and notes that the key barrier to response is less the sequencing of the virus and more the method of manufacture of vaccines.

Like the 2011 papers, the NSABB was consulted on the publication of the sequencing and then reconstruction of the 1918 flu virus. It eventually advised adding clarifying language around the benefits and risks of the work, which the report's authors did. But what was noted—then and six years later—was that the NSABB fundamentally lacked the power to do anything but ask for clarifying language. At the end of the day, any pause or hold on publishing the research was at the discretion of the scientific journals that had accepted the papers. And while the journals cooperated, they did not have to highlight the lack of true governance or power to prevent the publication of potentially dangerous research.[21]

Framing the Debate

The prehistory of GOF underscores a number of important issues. The first and most obvious is that while GOF describes a specific kind of experiment, it is lodged in a larger set of debates about risky life sciences research. Of the so-called experiments of concern devised by the authors of the Fink Report in the wake of the mousepox and polio synthesis experiments, GOF research falls into categories 1 (increasing virulence), 4 (transmissibility), and 5 (host range). GOF research is thus in one sense simply a subspecies of all dual-use research.

Next, debates about GOF research mirror those of dual-use research in how they understand the aims and methods scientists use. The mousepox study is indicative of this. It is not simply the case that GOF research *happens* to result in enhanced virulence, transmissibility, or host range. It is the *point* of that research. That is—and according to scientists who conducted the mousepox research, not policymakers—the experiments were *meant* to demonstrate that a cluster of novel mutations will result in a pandemic disease, often by discovering what those mutations are by engineering such a potential virus.

Note that this is not to say that GOF research, in view of the above, is simply unethical as a result. There might be reasons to do just what GOF intends. The point is that unlike some dual-use research in which a discovery might be unexpected, or regular "prospecting" for viruses in nature may happen to turn up a dangerous bug, the point of these experiments is to discover the path diseases take to pandemic disease. Like all science, this work might not be successful, but failure doesn't change the aims of a project.

The polio experiment gives us a useful counterexample to GOF. Unlike GOF, what made the synthesis of polio concerning was that it provided a method of creating a potentially enormous set of pathogens—already occurring in nature, but also potentially totally novel. It was a dual-use experiment whose product, polio, was not particularly dangerous in the moment but signaled a broader change in how biology is done. GOF research is not that: it often utilizes very old methods; serial passaging is a technique that biologists have used for much of the twentieth century and well before the genetic age. The direct products of GOF research are, almost always, what cause concern because they are novel viruses.

Finally, the synthesis and reconstruction of 1918 H1N1 influenza gives us an idea of what is entailed in GOF research. More often than not, when we say GOF, we are not referring to a single method in a single experiment. GOF research occurs as a series of experiments to develop novel viruses, much like resurrecting old ones. Suitable templates need to be found, for a start. Then the work of creating new viruses that are viable and have the relevant genetics takes its own time. Once these viruses are created, they need to be validated for their relevant properties and possibly optimized for their hosts. Only then is a virus fully complete.

Moving on, we can now look at contemporary GOF research. This will then give us an insight into the how and why of the controversy it has raised—and where critics and proponents alike overlooked details in the course of the debate. It will then set the stage for discussions of policy, ethics, and the future.

THE SCIENCE OF GAIN OF FUNCTION

GOF research is scientific research. That means that as a start, before we delve into GOF policy, we should look at what makes this particular research tick. In this chapter, we do just that.

It's important to note that increasing our understanding of what GOF research is doesn't *necessarily* give us a better idea about policy and ethics. Knowing which reagents are used in an assay doesn't tell us about the risks and benefits of research on its own. On the other hand, knowing more about the ferret, the animal used in GOF influenza experiments, may improve our knowledge in important ways. So we're going to examine the two most famous cases of GOF research in detail to understand the how and why of these experiments with an eye toward the rest of the story. I'll begin with the papers themselves, and then talk about the ferret and its role in GOF influenza research, and how different animal species in different kinds of GOF experiments might change how we think about GOF research. I'll finish by asking what kinds of questions GOF can answer uniquely, and then speak to the differences we might expect between exploring and engineering viruses.

The Structure of an Experiment

The first paper of interest is called "Airborne Transmission of Influenza A/H5N1 Virus between Ferrets"; its first author is Sander Herfst, and its final (and senior) author is Ron Fouchier, both with Erasmus University's department of virology.[1] The basic motivation for this work is that while most flu pandemics have been caused by reassortment—genetic mixing—there's no reason that a sneaky flu virus might not simply evolve, over time, the capacity to become a pandemic strain. Because H5N1 is one of the most concerning flu viruses when it comes to a potential pandemic, the team set out to see if an H5N1 virus could develop the relevant kinds of traits to cause a pandemic, with a focus on the big one for H5N1: transmissibility, bearing in mind that right now H5N1 is confined largely to birds, which is bad for nature, agriculture, and pet birds, but not *directly* bad for humans.

The team chose a virus called A/Indonesia/5/2005 (H5N1). There are several flu virus samples at play in these

Figure 2 Flu virus naming conventions.

two papers, so it's worth unpacking what kind of code that represents (see figure 2). The "HXNY" is the virus subtype. The "H" toward the end of the name is for hemagglutinin, the protein that influenza uses to detect and bind to proteins in the respiratory tract of its hosts, and then enter its cells. The number is the subtype, as is the number that comes after the "N"—neuraminidase, which allows the virus to exit the cell again. There are 18 Hs and 11 Ns, though the ones you'll hear most often in the context of humans are H1, H2, H3, and H5. H1N1 caused the 1918 pandemic and the 2009 "swine flu" pandemic. H2N2 is responsible for the pandemic of 1957–1958 that killed 1–1.5 million people. H3N2 caused the 1968–1969 pandemic that killed "only" up to a million people. And H5N1 is the "bird flu" that is the source of the controversy in the Herfst paper and has that dread 60 percent following it around.

The "A" in the name stands for influenza A, the virus type. Only influenza A viruses get the H-N naming system. There are three other genera of influenza virus, helpfully named B, C, and D. All of these viruses infect humans; influenza A viruses, however, are largely (but not exclusively) avian in origin, though they can emerge in ferrets, raccoons, tigers, and even whales, among other mammals. Influenza B viruses infect humans, pigs, and seals; C, only humans and pigs (of which we know), and D seems to favor cattle, humans, pigs, and horses.

The "Indonesia" is where the sample was isolated. Importantly, while we often hear blame for the emergence of diseases in Southeast Asia placed on things like "wet markets," that's only a small part of the story. As ecologists have noted, there's a much easier explanation for why pandemic viruses so frequently emerge in Southeast Asia, sub-Saharan Africa (e.g., Ebola), or Latin America (e.g., Zika). These are the regions with the highest biodiversity in the world—a phenomenon called the latitudinal biodiversity gradient. Basically, the closer you are to the equator, the more animals there are, and the more *kinds* of animals there are. And with that diversity comes a diversity of the things that live inside those animals and the chances a human will happen across one. There are simply fewer birds, and kinds of birds, the further you go from the equator and thus it's harder—but not impossible—to find bird flu in the wild.

The 5 is the strain number. These were sequential at one point, with numbering based on the number of strains from a particular flu genera, location, host species, subtype, and year scientists isolated the strain (the "2005"; in some cases, this is abbreviated to just the last 2 digits, such as 97 or 08). Lately, things have gotten more complicated as the number of samples has *exploded* and may sometimes include an identifier of the institution that has isolated the virus. So you might sometimes see "A/Michigan/UOM10045433565/2022 (H3N2)," an H3N2 strain of influenza A found in birds in Michigan that was sequenced by the University of Michigan (UOM) and appears to be one of many, many such viruses it's collected.

But for now, A/Indonesia/5/2005 (H5N1) is our dance partner.

The researchers first chose to find a version of their virus that they knew, if it emerged in nature, would bind to the upper respiratory tract of the ferret. A key roadblock for H5N1 is that it is designed to grab onto proteins that arise in birds—those in ferrets and humans are a little different, which as far as we know is one of the big barriers to a bird flu pandemic. The Erasmus team had, back in 2010, already shown a couple of key genetic changes: the N182K, Q222L, and G224S genes. The numbers reference the position of the amino acids these genes create on the H arm of the virus; remember that this is the arm that allows the virus to grab onto host cells. The team created a number of genetic mutants of A/Indonesia/5/2005 (H5N1), and then infected ferrets with either the wild version of the virus or one of these mutants. While the original virus was in fact the best inside the ferrets, the team found that the virus with both Q222L and G224S was the one that was best designed for human airways *and* replicated best once in the cell.

That was experiment 1 out of 7. Experiment 2 was to make sure it was a virus that really liked being in mammalian cells. This is where another gene comes in, encoding the PB2 protein. We've known about this one since 2009, and the researchers took their Q222L and G224S mutant and added in the gene that encodes an acid called E627K in PB2. The resulting virus, the *very* clunkily named A/H5N1$_{\text{HA Q222L, G224S PB2 E627K}}$, was inserted into ferrets. The researchers then sat those ferrets in cages next to uninfected ferrets, and . . . nothing happened. They had a virus that did fine *in* mammals, but that wasn't the point; they wanted a virus that jumped *between* ferrets.

It's important to stop and note that folks will sometimes assert that the above, just experiment 2, is GOF because genes are being inserted to "gain" function. But remember, we're talking about GOF *research*, a series of experiments that are part of the search for a particular set of properties in a microorganism. It's not just that *some* function changes, and certainly not just any function. We know

H5N1 can do okay in mammals; humans have died from it. What we want to see is that next step, transmission.

Enter experiment 3, with my emphasis: "Because the mutant virus harboring the E627K mutation in PB2 and Q222L and G224S in HA did not transmit in experiment 2, we designed an experiment *to force the virus to adapt* to replication in the mammalian respiratory tract and to select virus variants by repeated passage (10 passages in total) of the constructed A/H5N1$_{\text{HA Q222L,G224S PB2 E627K}}$ virus and A/H5N1$_{\text{wildtype}}$ virus in the ferret [upper respiratory tract]."[2] That is, the bits and pieces of what is called "reverse genetics," where scientists change an organism's genetic sequence to find out how its behavior changes, wasn't enough. The researchers now needed a little help from nature and in particular from a process called "serial passaging." As noted earlier, serial passaging is a process of taking a pathogen like a virus, inserting it into an organism, letting it percolate and mutate, and then extracting it and inserting it into another organism. This is done over and over, hence the "serial" part of the passaging. It is a venerable technique and responsible for, among other things, vaccine developments that have saved millions, including one of the polio vaccines. In the case of our ferrets, the passaging was intranasal—right up the nose. You shoot the virus up the nose of a ferret, wait four days (collecting swabs and data all the while), kill the ferret, scrape flu out of the nasal passages and lungs, and then shove that up the nose of the next ferret.

The third experiment generated a series of slightly modified viruses that all had small changes to their genetics. Experiment 4 tested whether this soup of modified viruses contained a version that could transmit by air. Collecting virus from the tenth ferret in their sequence, the researchers gave this sample to new ferrets, and the next day sat their cages down next to "naive"—that is, uninfected—ferrets. Ferrets near others infected with the original, wild virus didn't get sick. Those next to the results of the serial passaging did. Those viruses still contained the original changes the scientists had introduced to the genetics of H5N1, but with a few extra bits and pieces. Together, the experiment suggested that as few as five changes to the genetics of H5N1 would be needed to create something mammalian transmissible.

None of the ferrets in the fourth experiment died. In the fifth experiment, six ferrets were infected with a huge dose of the modified virus and were found dead three days later. It's important to note, however, that by huge dose we mean *a million times* more than you would ordinarily need to get sick. This huge dose is designed to test the ability of the virus to cause pneumonia by forcing a severe illness in an animal. What the researchers found is that if you give a ferret a huge dose of virus, it will die, but there's no death in this experiment for everyday, run-of-the-mill airborne transmission—there simply isn't enough virus to cause death.

Experiment 6 was a test of the new virus's sensitivity to oseltamivir, also known as Tamiflu, an antiviral drug given to people with flu. The virus turned out to be as sensitive as normal to Tamiflu and not resistant to our frontline treatment. This was followed by experiment 7, which tested the virus against antisera (the plural of antiserum) that are potential candidates for H5N1 vaccines. The researchers found that the HA arm—the one that matters for stopping a flu virus from getting into a cell—reacted just fine to the antisera and was even more sensitive to it than normal wild type virus.

The final experiment was a test against human sera to see if humans were innately immune to the modified virus. The answer was a resounding "no": humans don't have antibodies against this virus's HA arm, and thus are vulnerable to the virus should it emerge in the wild or otherwise be introduced into the human population.

So what did we find? It is possible for H5N1 highly pathogenic avian influenza to jump from birds to mammals in a serious way, without an extra virus to borrow genes from—and it is a small jump, requiring only 5 of the 327 amino acids in the hemagglutinin arm in order to get the virus to jump between ferrets, and by implication, humans. The virus doesn't seem dangerous in the way that wild H5N1 is, though it could still be quite dangerous relative to existing flu strains. It reacts to antivirals and potential vaccines quite well. But we are definitely vulnerable.

Recombination and Reassortment

What sets apart the next study—with first author Masaki Imai and senior author Yoshihiro Kawaoka, both at the University of Wisconsin at Madison—from the previous one is that unlike the Herfst paper, the research team led by Kawaoka was interested in what they considered the most likely natural strategy for H5N1 breaking into humans full time: reassortment.[3]

Reassortment occurs when multiple different viruses inhabit the same host and recombine into a novel form. It's how we got the 2009 flu pandemic, when three viruses met up in a pig: a H3N2 swine virus from North America; the HA (H1) segment in cows that has been around since it got into them during the 1918 pandemic; and the NA (N1) and another "M" segment from Eurasian flu species that is avian in nature but has been around in pigs awhile. (This M segment was later found to increase the activity of the NA segment, a "gene-gene" interaction that will become important later.) Flu viruses are legion, and even if they are not immediately dangerous, they can occasionally combine in the right fashion and develop a taste for the most dangerous game.

So the team led by Kawaoka started by noting that the segment of flu that really clinches the move to a new host is HA—again, it's the bit that gets the flu into you. This is similar to the insights of the team that produced the Herfst paper, but this time the researchers wanted to know if you could produce a virus that used the H5 HA segment to get into hosts.

To start, the team did something called "error-prone mutagenesis." It uses the technique of the polymerase chain reaction (PCR)—a technique the public is now familiar with thanks to COVID-19—but instead of creating reliable copies of RNA, the reaction is modified to create little mistakes in the copied RNA. This is usually done by changing the solution in which the copying reaction happens, and is a standard way of creating large numbers of slightly varied segments of RNA—ultimately a "library"— that can be screened for interesting properties. The mutations the team cared about were on the "globular head" of the HA segment, which is the bit that includes the piece of machinery that binds to host cells. This effectively created a library of new H5 HA segments from which the team could discover one that could get into mammalian cells.

The next step was to make these segments usable. The team combined its new HA segments into plasmids, which are little circular pieces of DNA that can copy their genes into other genetic machinery, along with an NA arm from a wild avian flu virus: A/Vietnam/1203/2004 (H5N1), which as you'll recall is influenza A H5N1, from Vietnam, the 1203rd sample from 2004. This little plasmid of H and N was inserted into a human embryonic kidney cell, a kind of staging ground for recombination, along with plasmids that had the last six segments of the flu, all from A/Puerto Rico/8/34 (H1N1), which is an H1N1 flu strain commonly used in flu research.

Which combinations worked? To figure this out, the team took red blood cells from a turkey. Turkey red blood cells are different from those of other birds in that they have equal numbers of the proteins that H5N1 viruses as well as human-infecting viruses like. (The western American magpie apparently has even more humanlike receptors, but is perhaps—said as a survivor of Australian magpie attacks—harder to catch and cultivate than the turkey.) Stripping the birdlike receptors out with salmonella, which has a special compound that eats sialic acids (the kind of receptors these are), gave a turkey cell that only reassortant viruses with the right, human-attacking proteins could grip onto. The library was then painted onto the cells as a film (called "adsorption," different from "absorption") and then washed so that all the weaklings fell off. From this, 370 viruses were found that bound to the relevant proteins on the turkey cells; of these, 9 started to clump the turkey cells together, a process called "agglutination," which is how flu scientists test for particularly active flu viruses as well as antibodies against flu viruses. All 9 had mutations—a mutation we've seen before (N186K), but also some others (called Q226L and S227N). These are all recognized to be involved in binding to human-type receptors.

Quality control led to dropping 1 of the 9 because it reverted to its wild form in the lab during testing, leaving the researchers to study how well the 8 remaining viruses bound to humanlike receptors. To do this, they compared the abilities of the 8 to a control virus that had the same basic structure as the A/Puerto Rico/8/34 (H1N1) virus used to create the modified viruses, with the H and N bits from A/Vietnam/1203/2004 (H5N1)—in essence, a reassortment of the starting position of the experiment, compared against the results of the mutations they induced. Of those 8, 1 ultimately bound *only* to humanlike receptors, with 4 others binding in greater or lesser degrees to both bird and human receptors. This is the one avian virus out of 2 *million* in the initial library that the researchers could bind to humanlike receptors as strongly as seasonal flu viruses. This is an elegant display of evolution in fast-forward—the random mutations of a new combination of viral genes, millions of times over, selecting for a single virus that had shifted from birds to human receptors.

Note I said "human *receptors*." Whether it worked in human airways was another question. So the scientists used lung and trachea tissue samples to check if the viruses could get in. Two did. The others, despite binding to humanlike receptors, didn't do it well enough to infect cells.

Enter the 2009 flu. The authors, having isolated the genes that produce the amino acids in the mutants they knew bound to human receptors, generated new viruses that both had combinations of those genes and were attached to a sample of flu virus from 2009: A/California/04/2009. Then they turned to ferrets to see if they could get the viruses to replicate. After some tweaking, they found that another mutation—called N158D—was needed to improve the replication of the virus in ferrets and their nasal turbinates, though not in their trachea. Put another way, the sniffles were the trick with this flu, less so coughs.

This is the point where the Imai and Herfst studies start to converge through serial passaging. Rather than doing this to just force their viruses to work in ferrets, however, the team at the University of Wisconsin at Madison first tested each virus to see if it transmitted at all. They found that only a few did, but they found an additional mutation had been picked up in one of them, called T318I. They took this virus and tried again, with success: four out of six ferrets turned up positive with this flu between three and seven days after hanging out next to their infected friends. This is almost exactly what we'd expect with something about as dangerous as the 1918 flu, and the researchers note that in unpublished data, they actually had confirmation of this transmission pattern.

After checking to make sure T318I wasn't doing all the work in this new flu (it wasn't), the story goes much the same: validation. The new virus was tested to see how well it fused to cells, and the result was that it was similar to wild flu viruses. The virus was then tested for its resistance to heat, and the researchers found it was considerably more stable in hot environments than its ancestors. They finished with testing how nasty the virus was, and found that while the modified flu caused weight loss and lung lesions in the ferrets, none died, and those symptoms that were found were less severe than wild viruses.

The Ferret Question

These are obviously complex research studies, consisting of multiple discrete experiments each designed to answer different subquestions in a larger project. Before turning to what that project *is*, however, we should talk first about what knowledge we can generate of any kind from the star of these shows: not the scientists, but their furry subjects.

The ferret has long been linked with the study of influenza. After an influenza epidemic in 1933, Wilson Smith and colleagues at the National Institute for Medical Research's Farm Laboratories in Mill Hill, outside London, collected throat washings from several patients, and proceeded to assay every available experimental animal in their collection to see if any of them would become infected. By a stroke of luck, a population of experimental ferrets was being maintained to study canine distemper virus, and two ferrets tested began to demonstrate all the classic symptoms of the flu: fever, food avoidance, weight loss, sneezing ("the ferret has an exquisite sneeze reflex," reported a researcher), fatigue, and a runny nose.[4] The work of Smith and colleagues, in the end, led to the first-ever isolation of the influenza virus and demonstration of its transmission through airborne particulates, which had been suspected but never confirmed.

What makes the ferret a particularly useful model is that it is the only known model that "can present both the pathogenic and transmissible features of influenza virus infection."[5] That is, ferrets will present a course of disease symptoms like humans, giving us an idea of whether a particular strain of influenza would lead to severe disease. But they also spread the disease via both droplet and aerosol transmission, giving us a way to determine whether a strain would be highly transmissible among humans. As these are the two variables that most determine the risk of a human influenza pandemic, the ferret model has become increasingly important.

All other model organisms for influenza miss at least one of these features. Mice require specially adapted influenza viruses and thus cannot be used to directly test the pathogenicity of viruses that could infect humans, and do not transmit influenza via aerosols or droplets. One strain of guinea pigs seems to mimic human transmission behavior, but does not exhibit clear signs of infection.

Philosophers of biology such as Michael Dietrich and colleagues have argued for a collection of twenty different features they suggest shape the choice of model organisms in the life sciences. These should not be interpreted as a "checklist" for a "good" model organism; rather, they are something more like an overlapping collection of virtues for which scientists typically argue when they defend the use of a model organism. Three of these features are commonly cited in scientific discussions of the ferret model in influenza research:

1. Phenomenal access to the relevant features of influenza: "in the sense of instantiating its typical features or providing insights that can be used towards understanding the phenomenon in question"

2. Translational potential: ferrets have relevant "physiological or genetic resemblance to humans"—a common sialic acid binding site in the upper respiratory tract, where influenza viruses most often infect humans, and where avian hosts have a binding site predominantly found in the human lower respiratory tract.

3. Responsiveness: ferrets offer better opportunities for the experimental manipulation of features of interest to researchers, primarily their respiratory anatomy, which

exhibits easy access to both upper- and lower-respiratory features.[6]

That is, ferrets not only give us valuable information about the world but also are easy for scientists to study in useful and interesting ways.

Yet in the ferret literature, scientists recognize (and lament!) that their systems fail to embody at least seven of the other virtues that are common in other model organism research subjects like mice. Ferrets frequently present difficulties for both ease of supply and financial costs, as initial costs are higher, standard equipment for the husbandry of ferrets is not widely available, inbred strains have not been developed, and influenza is occasionally already endemic in breeders' populations of ferrets. That is, ferrets are more expensive and trickier to handle than their murine cousins.

Animal ethical considerations are more significant in ferrets than many other animals used in research. It isn't that ferrets are a more ethically sensitive subject, but a lack of "best practices" that one might find in systems such as mice slows animal ethics committee approval. Put another way, the people who review ferret studies—who are usually a committee at a university—are used to mice, so they don't know what to look for in ferret research.

Standardization is often lacking, as highly inbred strains of ferrets have not been developed in the way that they have been in rodents. A trick with mouse research is that we know almost *exactly* what the genetics of each mouse is when we do experiments, but in ferrets there's an unknown amount of genetic diversity between populations that hampers the kinds of conclusions scientists can make in influenza research. The viability and durability of ferrets is frequently limited, as their expense often means that sample sizes of as few as five animals are regularly used, and stocks are not usually maintained by the laboratories performing the research. The availability of methods and techniques as well as the quantity of resources to interpret ferret experiments are underdeveloped too. On the practical side, viral inoculation methods, study end points, and necropsy methods are not necessarily shared across all labs. Ferret-specific reagents are often commercially unavailable. More broadly, the ferret genome was only published in 2014—notably, after the controversial GOF research discussed above was announced in late 2011.

In short, ferrets are a difficult, expensive, and minimally standardized system. The simple existence of scientific worry does not mean that ferrets somehow "fail" to be a model organism, however: rather, the overall summaries of the effectiveness of the ferret model are often quite nuanced, even those written by virology researchers themselves. Jessica Belser and colleagues write that "these confounders result in heterogeneity with regard to procedures and practices established at all levels of research, from individual investigators or institutions to broad country-specific regulations."[7] Ding Oh and Aeron Hurt write that

> the use of ferrets for influenza studies has been limited by factors such as animal availability, genetic heterogeneity (out-bred), the requirement of a complex husbandry facility and caging system, and a lack of immunological reagents and genetically modified mutants for immunological investigation.... Ideally, a larger number of ferrets should be used but limitations such as high experimental cost, low animal availability, limited caging capacity and ethical constraint, typically restricts most studies to group sizes of five or less ferrets.[8]

Ferrets are therefore, at the very least, a peculiar model organism—peculiar enough to justify an exploration of whether the kinds of limitations that ferret researchers have mentioned should lead us to reevaluate the role of ferrets in influenza research and the kinds of conclusions that are drawn from them. While remaining the only system capable of modeling both transmission and infection behaviors in influenza research, the scientific community itself recognizes that the system has a host of quirks and flaws. Of course, ferrets are not particularly unusual in this regard. Scientists will often have good reasons to use what might appear from the outside to be "better" and "worse" model organisms in different contexts, including—importantly—there being no better model available for a crucial set of research questions, as in influenza.

What Questions Does GOF Answer?

Given what we know about the canonical GOF studies and the animals the scientists use to create knowledge in the particular part of those studies we popularly recognize as distinctive of GOF, the next thing to ask is, If GOF is the answer, what is the question?

There is a lot of back and forth in the literature on GOF on what these studies *ultimately* do for humanity. The three answers commonly given are:

1. Alerts the community about the dangers of H5N1
2. Enables disease surveillance
3. Enables vaccine production

These are fairly strong claims, and the degree to which they are true among so many other things being done to achieve these goals is frequently hard to substantiate. A better place to start, however, is to ask what the scientists, in their writing, thought of the question GOF was meant to answer. The Herfst paper is clear in the abstract: "Highly pathogenic avian influenza A/H5N1 virus can cause morbidity and mortality in humans but thus far has not acquired the ability to be transmitted by aerosol or respiratory droplet ('airborne transmission') between humans. To address the concern that the virus could acquire this ability under natural conditions, we genetically modified A/H5N1 virus by site-directed mutagenesis and subsequent serial passage in ferrets."[9] The Imai paper likewise makes early claims in its abstract and aims, but is even more definitive in its concluding passage: "Our research answers a fundamental question in influenza research: can H5-HA-possessing viruses support transmission in mammals?"[10] On first blush, these questions seem to be answered definitively and in the positive. The Herfst paper shows that through select mutations and passaging, A/H5N1 can develop the ability to transmit between mammals. And the Imai paper answers the question of whether influenza A viruses with H5 hemagglutinin arms can transmit between mammals.

But note some wrinkles. It's not clear from the previous section that showing something in ferrets is the same as showing it in humans. This makes the concern in the opening sentences of the Herfst paper a little less apparent than it seems to be. In fact, scientists in 2012 dismissed security concerns precisely because ferrets aren't humans.

This may strike the reader as scientists attempting to have their cake and eat it too. The strength of the connection between ferrets as an experimental organism and humans as its target motivates the experiment, but that connection is not strong enough to motivate fears that the results of the studies are directly transferable to humans. Rather than a mere rhetorical device, however, this tension points to deeper questions about flu studies and the role of ferrets within them.

While we'll see a little more of the politics of science in the coming chapters, there is a charitable response. And that is, as we saw in the last section, that there's just no better, more ethical way to figure these questions out. While we do perform so-called challenge studies in influenza research, these are on mild seasonal strains as opposed to potentially killer pandemic strains. Other than infecting humans with genetically modified flu on purpose, ferrets are our best shot. And short of actually seeing a pandemic virus in the wild—at which point it might be a little late—we're stuck with the lab to sort these things out.

Of course, this doesn't come close to the kinds of ultimate aims we started with. The first one, alerting the community, is almost hilarious in its circularity. As scientists like to remind people, the kinds of experiments we're talking about take money, resources, and training. And all of that has to come from somewhere: while Fouchier and Kawaoka I'm sure do just fine as high-level scientists at major research institutions, they're unlikely to be moneyed scientists, self-funding their work in the fashion of the early scientific revolution. No, these studies were commissioned after lengthy grant proposals submitted to government funding agencies. The finances for the Herfst paper come from the National Institutes for Allergens and Infectious Disease in the United States as well as the NIH director's Pioneer Award that supports scientists in doing high-risk, high-reward research. They also received funding from the Dutch government and the European Union along with support from other government-funded research institutions. The Imai paper is much the same, albeit with different grant names and locations.

If there's a need for an alert, then where on earth is this money coming from? It's true that the world was, is, and continues to be woefully underprepared for pandemic disease. How much worse would COVID have been if it had traveled faster or killed even just a slightly higher proportion of people? A 0.1 percent increase in mortality on a billion people is still a million extra dead. But it's just not clear that the world needed an alert in the form of a scientific paper so much as an alert in the form of a protest on Capitol Hill in the United States, or outside the WHO in Geneva.

Does this research enable surveillance? We are attempting, essentially, to predict the evolutionary future because doing so is instrumentally valuable to preventing a catastrophic disease pandemic. What kinds of mutations might plausibly arise within a given viral lineage, and how might those mutations lead to changes in the viruses' ability to infect and transmit between humans? In several cases, we have successfully made such predictions, testing, for instance, certain kinds of mutations in GOF contexts that were later discovered in human influenza viruses, which went on to be incorporated into the influenza vaccine. Any critique of this research must, to be sure, take this success into account.

Nonetheless, the kinds of concerns that we have already seen with the methodologies adopted in and the statistical power of ferret-based research remain important sources of uncertainty. We can see this in at least three different ways. First, results arising from a GOF research program require an extensive amount of comparison against baselines; we are fundamentally in pursuit of enhanced virulence or transmission with respect to some external standard. But our capacity to compare results across laboratories is precisely one of the difficulties of working on ferrets: given the differences between ferret populations along with the myriad ways to inoculate ferrets with an influenza virus, measure viral end points, and detect transmission, being certain that such a comparison of viral characteristics is legitimate is not easy.

Second, such uncertainty is magnified by the small scale of ferret experiments. Sample sizes in the single or low double digits—ten in the fourth experiment in the Herfst paper, and six in the fifth—especially when sampling against a population with an unknown amount of genetic diversity in the absence of standardized, inbred strains adds extra uncertainty to research results.

Finally, in the context of many GOF research programs, these uncertain results are magnified yet further by the overall experimental design. GOF experiments like those in avian influenza discussed above turn not only on the low-level results in ferrets but also on a number of other uncertain experimental choices: we must select a strain of avian influenza on which to work (among the many known and likely many more unknown such strains), and we often target a specific site at which to mutate that strain (sometimes drawn from prior experience with other influenza viruses, and sometimes from our biochemical knowledge of relevant protein binding). Both are, in essence, gambles on the correct initial conditions for our future evolutionary predictions and magnify the error bars (so to speak) on our final conclusions.

Finally, when it comes to vaccines, it has long been recognized—back to the resurrection of 1918 influenzas at least—that knowing which virus to select for in making a vaccine isn't the slow bit in vaccine production. Rather, it is the manufacturing process that makes things slow. And further, making pandemic flu vaccines for viruses that do not and may never exist in nature is an expensive task that is infeasible for most governments, and unpopular for the rest. This process may be made faster for mRNA vaccines, but validation, storage, and distribution are still heavier burdens than knowing one of the paths H5N1 might take into our lungs, hearts, and brains.

This is not to say this research isn't or can't be useful. But it is noteworthy that these experiments are tiny pieces in a larger puzzle of preparing for disease pandemics. The size of their role, in different contexts, has determined how they have interacted with policy and the story of what came next.

Engineering versus Exploring Viruses

Let me offer coda to this inquiry into the science of GOF. It is worth noting that, as I described last chapter, GOF is an intentional process. It is not that scientists have tinkered to try to understand mutations in viruses and simply *happened* to find a virus that had pandemic potential. They were looking at and screening viruses—in some cases, millions—for the right combinations of genes that would signal pandemic potential.

But this makes the processes, we can note, somewhat different than what we often consider "science." The questions here are less about estimating potential and more about identifying possibility—that nature, with enough force, can be engineered to result in pandemic influenza. Nevertheless, this term, "engineering," carries significantly different implications than science, despite being used in similar contexts.

Engineering as a process often tends to be cyclic around a particular need or constraint that engineers are attempting to overcome. To get at this need, engineers will research and then select a solution that is tractable and prototype a system that will accomplish their goal. They'll then test and evaluate the prototype before improving or *optimizing* the prototype into a final product. There are lots of different "engineering cycles" to be found in professional societies, classrooms, and firms, but they all typically involve a system of prototyping and optimization.

Both of these papers, then, look more like engineering in some ways than science. The Herfst paper doesn't even ask a question! Instead, the aim of these researchers was to create a mammalian transmissible HA H5N1 virus, and

These experiments are as much about discovering mechanisms and showing how evolution works as they are about the more concrete public health goals ascribed to them.

they selected their methods based on various constraints and opportunities provided to them by the science of the time. They created a prototype or a library of potential prototypes, and then optimized their results through serial passaging and vetted them against a range of tests to put the viruses through their paces.

Why does this matter? Well, because these studies might be less about generating public health knowledge, aside from proving that these things *can* happen. Rather, they are a set of pivotal pieces of research using methods to create and optimize viruses for particular functions. These functions tell us something about the world, yes. But their genetic information is less useful for predicting a viral emergence than it is for deepening our understanding of which bits of the flu genome do what.

In this, the Imai paper is deeply reflective in its closing sentence: "Although a pandemic H5N1 virus may not possess the amino acid changes identified in our study, the findings described here will advance our understanding of the mechanisms and evolutionary pathways that contribute to avian influenza virus transmission in mammals."[11] That is, these experiments are as much about discovering mechanisms and showing how evolution works as they are about the more concrete public health goals ascribed to them. To be crystal clear: this kind of virology is important work. Yet when decoupled from their public health aims and the convoluted path these studies might take to

This kind of virology is important work. Yet when decoupled from their public health aims and the convoluted path these studies might take to impact public health, they are more about basic influenza virology than they are about saving the planet.

impact public health, they are more about basic influenza virology than they are about saving the planet. In many ways, that's fine—it's a peculiar kind of logic that requires science to be immediately and obviously beneficial to more pragmatic aims. Still, it's worth acknowledging as we move into the debate about what to do when those broader aims toward knowledge conflict with health and human security.

4

THE EMERGENCE OF GAIN OF FUNCTION AS A POLICY DEBATE

It is worth noting that nothing about the last chapter really points to why these experiments received such a response from the US government and indeed the world. That is, we've just described the science and some of the important details of the science, but nothing about the risks, benefits, politics, or ethics of GOF research that make it so controversial. In the *next* chapter, the full consequences of the controversy will be laid out—almost a decade of policy debate involving professional societies, individual scientists, governments, and even the UN system. But in this current chapter, we're going to describe the big bang of the GOF debate in the form of the immediate response to our two paradigm cases.

The Most Dangerous Virus

Let's go back to November 23, 2011. *Science Insider*—the news arm of *Science*, one of the most prestigious scientific journals in the world—released an article with the headline "Scientists Brace for Media Storm around Controversial Flu Studies." The topic of the article covered the pending publication of two manuscripts in *Science* and *Nature*—the two studies described in the previous chapter. As journalists are wont to do, the stakes of the problem were laid out in poetic detail: "a man-made flu virus that could change world history if it were ever set free."[1]

At that stage, the results of the study had actually been seen by a fairly large number of people in the scientific community or at least the community that is historically interested in such findings. Fouchier had presented his team's paper earlier that month at the European Scientific Working Group on Influenza (ESWI) in Malta, where he reported how his team had "mutated the hell out of H5N1" to portray the first two experiments of the study. He then said that was when "someone finally convinced me to do something really, really stupid" by serial passaging experiments that we all know from experiment 4 of the study. The ESWI is the largest influenza-specific infectious disease conference on the planet, and so we can assume that, even though one reporter noted that Fouchier's session was on during the dread breakfast session at the conference, hundreds of scientists, public health practitioners, and policymakers already knew of the paper by the time *Science* hit the presses.[2]

By then, the paper and its counterpart by the team at the University of Wisconsin at Madison had been subject to deliberation by the US NSABB. That board's chair, Paul Keim, an anthrax expert, claimed, "I can't think of another pathogenic organism that is as scary as this one."[3] A month later, the *New York Times* would report that "for the first time ever, a government advisory board is asking a scientific journal not to publish details of certain biomedical experiments, for fear that the information could be used by terrorists to create deadly viruses and touch off epidemics."[4] It's worth noting that this is only partly true. In 2005, the Department of Health and Human Services (HHS)—just as the NSABB was being set up—had asked the *Proceedings of the National Academies of Science* journal to withhold publishing a paper modeling an attack by terrorists using botulinum toxin to poison the US milk supply.[5] And in 1979, while not a peer-reviewed scientific publication, *Progressive* magazine faced an attempt by the US government to censor the outlet from publishing an article on the construction of a hydrogen fusion bomb—an effort abandoned ultimately on appeal.[6] But this was certainly the first time a federal *advisory committee*, created expressly for the purpose of thinking about the security implications of the life sciences, had reached the conclusion that a paper should not be published.

The Request Not to Publish: Science versus Security

The request not to publish applied to specific sections of the manuscripts. In particular, the NSABB advised HHS to request of *Science* and *Nature* that the methods of the papers and other details that could "enable the replication of the experiments by those who could seek to do harm" be removed.[7] The NSABB also recommended that language be added to the manuscripts to better explain the goals and potential public health benefits of the research, and to detail the extensive safety and security measures taken to protect laboratory workers and the public. The leadership at HHS concurred with NSABB's assessment, and the request was passed to *Nature* and *Science*.

The basic reason given is the idea of a "blueprint"—which is the same kind of concern that was raised in 2005 in the aforementioned botulinum toxin study. That is, the studies were conceived of not as discoveries so much as *recipes* for the creation of novel strains of influenza. That is, someone with the right means and motive could use the research to create a human-transmissible strain of influenza A H5N1, and then spread it into the community and ultimately the world. The WHO, in late December 2011, added its opinion that the Pandemic Influenza Preparedness Framework, which had only entered into force that May, might be undermined if the scientists using samples negotiated through that framework did not conduct their research "only after all important public health risks and benefits have been identified and reviewed."[8]

While the precise details of those discussions are not available in the open record, the NSABB published a paper in *Science* on February 10, 2012. It noted that the discovery of a path by which influenza A/H5N1 could acquire the capacity for mammal-to-mammal transmission was an important step and that now "society can take steps globally to prepare for when nature might generate such a virus spontaneously."[9] This kind of benefit mirrors, almost exactly, the kind of questions we discussed in the previous chapter, albeit the one that might raise the most eyebrows. After all, avian flu had been on the world's radar since 2003. By 2005, Anthony Fauci, director of the National Institute of Allergy and Infectious Diseases and thus one of the chief funders of GOF research, had claimed that the world was almost 40 years overdue for a devastating influenza pandemic.[10] By the time the GOF papers came along, the United States had spent a little over $13 million *per case* of H5N1 attempting to head off a pandemic.[11] Scientists may have confirmed, through GOF research, that this money was, if not well spent, then at least not wasted. But as is common in security circles, including so-called health security, the possibility of a threat is not enough for preparedness. And interestingly, the Kawaoka finding gave us the beginning of an estimate: of two million mutants, only eight could bind to human receptors in the upper respiratory tract, and only one exclusively did so.

The NSABB then asked: Could the knowledge, in the wrong hands, allow for the construction of a genetically engineered virus as bad or worse than the 1918 pandemic strain? The NSABB hedged by noting that the scientists were well intentioned, trained, and scientifically rigorous. Yet they arrived at the conclusion that the potential risk of public harm was of "an unusually high magnitude," and on that basis recommended the modification of the research. They claimed further that, as scientists, they had a "primary responsibility 'to do no harm,'" echoing the common understanding of the Hippocratic oath. They concluded by remarking that this case had parallels to the 1975 Asilomar Conference on recombinant DNA—though they did not, unfortunately, mention that the conference had little to nothing to do with security, and everything to do with accidental or premature release.[12]

Lifesaving Research

The response from the scientific community was, to say the least, mixed. Some scientists praised the move, but many others were vehemently opposed. Writing in early January 2012, Peter Palese, then chair of the Department of Microbiology at the Mount Sinai School of Medicine in New York, wrote that publishing the studies without their crucial details was "akin to censorship, and counter to science, progress, and public health."[13] Palese pioneered reverse genetics, the kind used in the studies, and was also one of the scientists who worked on the 1918 influenza sequencing a reconstruction. He bolstered his claim in 2012 with a callback to that work in 2005, noting that the papers that followed the reconstruction of the 1918 flu led to discoveries that the virus was sensitive to the seasonal flu vaccine and available antivirals. He later claimed in the same piece that these discoveries obviated the possibility of a worst-case scenario of a rerelease of 1918. I suspect this last claim might now appear naive given what we have seen regarding vaccination, public sentiment, and political will during COVID-19; the presence of scientific knowledge and the will to turn that into public health reaction are different things entirely.

The scientific community, however, relented in part. In late January, a group of scientists asserted in a *Science* article that a 60-day moratorium would go into effect for work on avian flu transmission, like the Herfst and Imai studies.[14] That letter included Palese; Kawaoka, and Fouchier (who was the lead author); David Morens and Jeffery Taubenberger of the 1918 synthesis fame; Subbarao, who I spoke about in the first chapter; and other luminaries of the influenza community. The letter disagreed with the NSABB's conclusions, claiming that despite the public health benefits of the research, the "perceived" fear of the "escape" of the novel viruses had generated public debate. Recognizing the need to "clearly explain the benefits of this important research," the authors agreed to a 60-day pause in research to engage in discussion with the public, regulators, and others.

A brief detour here is worthwhile before we see the conclusion and the NSABB's change of heart. When scientists, and especially elite scientists, talk about "public discussion," we should always—to borrow a turn from my colleagues in anthropology and sociology—ask, "Who is this public you speak of?" It is clear now, but it was also clear then, that the statistically average person in the United States, much less the world, was not the public the scientists had in mind. To my knowledge, not even *60 Minutes*, one of the most watched general interest news programs in the world, covered this issue. The *New York Times* did, in a series of pieces by William J. Broad. Some smaller publications such as *Wired* and *Scientific American* covered the issue. And, of course, *Science* and *Nature* did so in their news columns.

But this is not a "public" debate in any real sense. The *New York Times* has 9 million subscribers, just under 3 percent of the US population, and they tend to skew younger and more educated than the general public. There were no public forums larger than this, and indeed even then the stories were far from front-page news. To say that *Science* is a public outlet in a broad sense is outrageous! While much of this content was published online, moreover, that's not to suggest that there was significant uptake from the wider public.

Rather, the "public" the designers of the moratorium were pointing at were the NSABB itself along with affiliated policymakers and scientists within and outside the United States. That is, the public debate *they* needed to happen was not one between voting citizens as to how their tax dollars should be spent when it came to scientific funding, what kinds of priorities were worth having, and so on. (It is not clear that a model of straight democratic engagement is *the* model we should choose either, to be clear.) Rather, this can be seen as a strategy between scientists and policymakers as to who could make the most persuasive and highest-pressure argument regarding the publication, or not, of the work directly to the people who were in power. And that persuasion and pressure had its effects in 2012.

A Change of Heart

On February 16 and 17, 2012, the "public debate" happened in Geneva at the WHO. To reiterate, and for those who haven't been, the WHO is hardly "public" in any broad sense, unless you can fly to Geneva, find accommodation in that famously expensive city, and take bus route 8 to the top of the hill that the WHO (or the OMS, to use the French abbreviation that appears on transit maps of the city) perches on. Nonetheless, this is the "public" the authors of the moratorium, then entering its fifth week, had in mind when they published their letter.[15]

The meeting was partly theater: Fouchier and Kawaoka passed out copies of their papers as they were reviewed in *Science* and *Nature*, and second copies redacted along the lines that the NSABB had suggested. Participants at the workshop signed a receipt of the numbered copies of the manuscripts (of which, recall, one had been publicly presented at the ESWI the year before) and returned them at the end of the meeting. The copies were shredded in front of the assembled workshop—surely a scientist's idea of what national security looks like, but tailored to the participants at that.

The resulting recommendation opposed the NSABB and suggested full publication. WHO assistant director general Keiji Fukuda noted that the potential public health benefits of the research generated strong support for publication and that the logistics of a secure approach to dissemination would be "very difficult to do overnight, if not impossible"—perhaps too quickly eliding the possibility that doing this "overnight" was itself a question that might need its own answer.

Keim, NSABB board chair, opposed the recommendation. And he did so as a kind of concern with conflict of interests, remarking that the workshop was almost exclusively made up of flu researchers. Keim's job as an anthrax researcher brought him much closer to the US national security establishment than influenza researchers typically find themselves; Fauci, in attendance and a representative of HHS, also disagreed with the workshop's conclusions.

Nonetheless, the scene change was in play. Even before the NSABB's next meeting, interviews with its membership started to show changes of heart. Arturo Casadevall, one of the inaugural members of the NSABB, claimed that he was "going to go to the next meeting with an open mind and listen to everything. This process is supposed to be deliberative. And this process is one in which you can think through it and change your mind."[16] On March 29 and 30, the NSABB met with Kawaoka and Fouchier, and received revised papers that clarified certain kinds of findings not immediately accessible in the accepted drafts that initially caused controversy—importantly for the virologists, that the mutant viruses appeared to be less pathogenic than wild avian influenza A. The Kawaoka paper received a unanimous approval to publish, and the Fouchier one was recommended for full publication on a 12–6 vote. The director of the NIH, Francis Collins, would endorse these recommendations on April 19 in a blog post; in that same post, the director of HHS, Kathleen Sibelius, is said to have concurred with that decision.[17]

The publication was not as straightforward as it seemed, at least for Fouchier, who also had to wrestle with Dutch export control laws for his publication. This is the one wrinkle in the story, and though it was ultimately resolved, it signaled an important difference between the Netherlands and United States in their processes. The NSABB is a federal advisory committee; it has no statutory authority to place *binding* recommendations on the agency that houses it, HHS. Nor does HHS have the ability to do more than *ask* journals to withhold publications in part or full. During the 1918 influenza sequencing controversy mentioned two chapters ago, the editor in chief at *Science* published an editorial noting that even if the NSABB had ultimately recommended against the publication of the reconstruction (which it did not), he would have done it *anyway*.[18] Between the first amendment in the US Constitution, the private versus public status of journal publishers, and a lack of any binding policy around dual use in 2012, there was nothing else to be done in the end—especially with an international endorsement for publication through the WHO workshop, which, while not an official UN decision, carried with it the legitimacy of that institution.

In any other debate, I suspect we'd say the scientists called the bluff of the NSABB, and the NSABB folded. We tend to phrase this instead as the members of that committee "reconsidering the evidence," but I think we should not underestimate the politics in science or the skill with which the scientists in question acted. All of that, however, was about to change, and it is important to note that the seeds of the conspiracies that would emerge during COVID were arguably planted not in labs in Wuhan, as goes those stories, but rather in meeting rooms in Geneva and Bethesda, where the politics and *esteem* of science often get made.

5

GAIN-OF-FUNCTION POLICY

In the beginning, I noted that the phrase "gain of function" is, in our discussion, a policy term rather than a scientific one. That is, as opposed to capturing a particular technique, it captures a set of studies that result in a kind of virus we are worried about. And while this "we" should probably be "the whole world," the reality for the most part is that it often means a small group of policymakers, activists, scholars, and professionals that stands in for the public in a niche policy debate.

And make no mistake, GOF, at least until recently, is niche. *Very* niche. Dual-use research is a niche policy area, comprising a cluster of policies at the national and international level that deal with science and security. GOF is even smaller. And within that it has to compete with wars in eastern Europe and the Middle East, climate change, inflation, hurricanes, overdose crises, police brutality, gun violence, and—of course—naturally occurring global pandemics. There's a lot to be worried about, even if you're someone like me for whom GOF is a core part of their career.

Yet policy has and continues to move. In the decade since the Imai and Herfst papers first caused a stir, a number of policies have emerged that have described different kinds of limitations and guidance for the funding, conduct, and communication of GOF studies. These policies are what we'll cover in this chapter. But first, two quick caveats.

For starters, these policies aren't the only ones that people conducting GOF research are covered by. Life scientists are, like all other scientists, enmeshed in a web of different kinds of rules and regulations. Some of these, like export control laws, pertain to national security and preventing the wrong people from getting access to critical technology. Others pertain to materials transfer and patents, which work to make sure that particular people get rich off invention and innovation. Then there is laboratory biosafety, which concerns the infrastructure and practices that stop scientists from getting sick from the diseases they study or releasing them into the community by accident. There are also animal welfare laws and ethics regulations governing research with humans (which GOF experiments don't use, but which GOF researchers might need to comply with in other experiments). There are international frameworks like the Biological Weapons Convention, International Health Regulations, and Pandemic Influenza Preparedness Framework, where states agree to create certain laws domestically or coordinate internationally with each other. And so on.

It's complicated, and it's no wonder that newcomers to the world of GOF often misunderstand the significance of particular kinds of research or "discover problems" that have been known for years. It is the work of entire careers to understand the way that biology is managed. And adding or removing any one of these things is like a game of policy *Jenga*, where players take turns removing pieces from a tower made of wooden blocks and stacking them back on top.

Second, at least in the United States, no specific piece of explicit GOF policy is a law. No piece of GOF policy is even a federal rule—a kind of policy in the United States that obligates interactions with the federal government to follow particular guidelines. Rather, these policies typically take the form of agency guidance for the review of research when deciding to fund scientists and then for the funding itself. There are no doubt several reasons why this is the case, but the most important one is that decision-making power over GOF is retained by the agencies that fund the research in the executive branch of the US government. Laws would require congressional involvement and invite legal challenges in courts. Rules follow their

At least in the United States, no specific piece of explicit GOF policy is a law.

own process of soliciting comments and deliberating policy, and can be extremely time-consuming. But all of this is to say, these are tiny changes, operating at their broadest at the level of the White House and at their most narrow within particular subdivisions of the NIH.

This is not the case in other countries, which—while not necessarily pursuing GOF policy per se—may have wrapped the issue of dual use into their national rules and laws. I suspect one reason for this is that in nations like the Netherlands, United Kingdom, Japan, and Australia, all of which have progressive policies on dual use, there is not a clean separation between the executive and legislature. This "fusion" between these two branches of government means, on the one hand, that engaging with parliament is unavoidable, as ministers are frequently engaged in the leadership of executive agencies; you can't keep a decision to the executive branch in principle. And on the other hand the policymaking process is, for better or worse, less intractable than it might be in nations like the United States.

A related issue is that many nations, unlike the United States, have a public service in which civil servants are not necessarily subject matter experts outside their roles in government. While big names in US science who are also civil servants—like Fauci and Collins—do exist elsewhere, they do not wield nearly the same formal and informal power in other nations. One consequence of this, I suspect, is that the politics of dual use and GOF are not as informed by the identities of civil servants *as scientists*. This has its drawbacks to be sure: if civil servants have no skin in the game of the science they regulate, they may be overzealous or misunderstand the stakes of what they are undertaking. But is helps, conversely, when you aren't colleagues with the same people you seek to regulate.

If it sounds like I'm hedging, you're right. This isn't a book on political theory, and so the history and justification for the architecture of entire national governments are beyond the scope of analysis here. It's simply not possible to say here whether the architecture, and unique position Fauci held in the NIH as one of the most prominent and powerful scientific voices in the United States, is a good thing or not, and whether his influence in the United States relative to someone like Anne Kelso, the CEO of the Australian National Health and Medical Research Council, is justified. But they are important insights that help frame the power these policies wield over scientists and their limits.

The First Policy

The first GOF policy emerged shortly after the initial Imai and Herfst papers were released. The policy, produced in early 2013, was brief and to the point, with a title indicative of how specific it was to the moment: "A Framework for Guiding U.S. Department of Health and Human Services Funding Decisions about Research Proposals with the Potential for Generating Highly Pathogenic Avian Influenza H5N1 Viruses That Are Transmissible among Mammals by Respiratory Droplets."[1] That is, the policy wasn't about GOF research writ large. It was just about the specific kinds of GOF research that the Imai and Herfst papers represented: studies that conferred mammalian transmission on highly pathogenic avian influenza H5N1. In 2013, they would expand this to include H7N9, which emerged as a lethal yet fortunately not terribly transmissible virus in China.[2]

The policy established an additional layer of review for funders of GOF research, which, at that time, was the NIH and only the NIH. After standard scientific review, any study "reasonably anticipated to generate a HPAI H5N1 virus that is transmissible between mammals by respiratory droplets" would be submitted for additional review. This review would consider seven criteria:

1. Whether the virus anticipated to be generated could be produced through a natural evolutionary process;

2. If the research addresses a scientific question with high significance to public health;

3. There are no feasible alternative methods to address the same scientific question in a manner that poses less risk than does the proposed approach;

4. Biosafety risks to laboratory workers and the public can be sufficiently mitigated and managed;

5. Biosecurity risks can be sufficiently mitigated and managed;

6. The research information is anticipated to be broadly shared in order to realize its potential benefits to global health; and

7. The research will be supported through funding mechanisms that facilitate appropriate oversight of the conduct and communication of the research.[3]

If the answer to all seven was "yes," then a second, departmental review would establish that the work in question had been appropriately assessed for risk, benefit, and meeting "critical needs" in the portfolio of research at HHS. This review would involve convening a group led by the Office of the Assistant Secretary for Preparedness and Response, the office in that department that deals with disasters and public health emergencies. That group would review the funding agency determination, provide its own input, and determine where the experiment fell not just within department priorities but also across the

US federal agency landscape, sometimes known as the "interagency." If the Office of the Assistant Secretary for Preparedness and Response issued a recommendation to move forward with the project, then it would be up to the funding agency to give the final "yes" to the project.

The entire framework ends with something of a warning. It says that investigators who submit proposals that fall within the scope of the framework are encouraged to be mindful of the seven criteria and "develop their proposals accordingly." That is, the criteria above are not simply a passive review mechanism but rather guidance for proposers of research to ensure they hit the right notes in their proposals.

It is important to keep in mind that the policy is limited, considering a very, *very* small set of virology studies. Not simply "only" flu studies, but only HPAI H5N1 studies. And not studies that change virulence or transmissibility, but only those that change the host range of the virus. Nonetheless, this is a policy developed over a year by a small group of people to respond to a particular crisis of trust and legitimacy brought about by the Imai and Herfst papers.

The Move from Security to Safety

The first policy still explicitly dealt with GOF research as dual-use research—research that could conceivably be *misused* by bad actors. In 2014, however, this framing of GOF started to change. That change would come in the move from security to safety.

Small groups within the scientific and policy community had raised concerns about laboratory releases for some time. In 2008, for example, the US Government Accountability Office had released a report stating that despite moves to the contrary—and in particular the proposal to move the aging Plum Island Animal Disease Center to the National Bio and Agro-Defense Facility in Manhattan, Kansas—there was no safe way to conduct research on foot-and-mouth disease on the US mainland.[4] In 2012, Marc Lipsitch and Barry Bloom at Harvard University wrote an editorial in the journal of the American Society for Microbiology, *mBio*, advocating explicit risk assessments for all research conducted involving "potential pandemic pathogens" (PPPs).[5] This term was coined by security researchers Lynn Klotz and Edward Sylvester some months before, and would emerge as a way of further framing the GOF debate by linking GOF research to the creation of these PPPs—later referred to in GOF policy as *enhanced* PPPs (ePPPs) to distinguish them from naturally arising PPPs, and then finally in 2024 as PEPPs, or pathogens of enhanced pandemic potential.[6] One of the great stories of GOF is that the US Department of Health and Human Services, unlike the US Department of Defense, really struggles with an acronym.

But the inciting reason for the change came about through three high-profile laboratory mishaps in early 2014. I say mishaps because only one of them is a "lab leak" in any real sense. The first mishap was the revelation that CDC scientists were cutting corners in working with anthrax, not taking appropriate steps to inactivate the bacterium before working with it or moving it between labs. Up to 74 researchers were thought to have been exposed to live anthrax, though none were found to have been infected. Most colorfully, descriptions of shortcuts like moving petri dishes of anthrax in ziplock bags emerged, painting CDC researchers as cavalier at best and reckless at worst.[7]

The second mishap and the actual "leak" occurred when the US Department of Agriculture started losing chickens. The researchers had ordered samples of low pathogenicity avian influenza to study in chickens. But when the birds in their labs were inoculated with the samples, the chickens began to die. It was found that the samples were not just low pathogenicity; there was *also* highly pathogenic avian influenza contaminating the samples. The news headlines read of the transport of this dangerous form of avian influenza—recalling that wild HPAIs don't transmit between humans—in FedEx trucks.[8]

The final mishap was the discovery of a box of glass vials in a fridge at an old Food and Drug Administration office in Bethesda. Of the many vials discovered, some were found to contain smallpox spores. The disease, eradicated in 1980, was previously thought to only be found in freezers in the CDC in Atlanta, and in its counterpart in Russia, the Vector Institute. This discovery was a biosecurity expert's worst nightmare but, in the context of the events of 2014, was framed as the hazard of the proliferation of risky biological research worldwide.[9]

This "proliferation" is not of weapons but instead typically cast as that of *labs*. A *Guardian* article by Ian Sample that year highlighted a string of laboratory accidents in the United Kingdom but, in conjunction with work by Alison Young at *USA Today*, ultimately centered on the growth in the number of high-containment laboratories.[10] These laboratories, designated biological safety level 4 or BSL-4, the highest of four possible biosafety levels according to the US manual *Biosafety in Microbiological and Biomedical Laboratories (BMBL)*, are often said to be home to the world's most dangerous pathogens.[11] What is probably more accurate is that—with the exception of smallpox, which doesn't occur in the wild anymore—they are home to the *research* on the world's most dangerous pathogens, and where scientists attempt to better understand those pathogens with the aim of better protecting against them.

The move from safety to security is significant for two central reasons. From the perspective of some advocates, like Lipsitch, gauging security risks requires special as well as often classified expertise and information. Safety risks,

"Proliferation" is not of weapons but instead typically cast as that of *labs*.

on the other hand, are arguably easier in theory to quantify without secret knowledge, though evidence on the efficacy of modern biological safety equipment also largely doesn't exist. Nonetheless, a sufficiently motivated funder like the NIH *could* bring that data into existence in a way that it can't with security, which may remain secret, but also relies on assumptions about what adversaries will or won't do in response to certain defensive measures.

But more important for moving the policy along was that the move to safety brought the issue home to the United States. It's one thing to note that faraway terrorists or foreign governments could use well-intended scientific research to create weapons. The safety issue and the possibility that a GOF experiment could simply walk out the door of a lab turned the moral stakes of the debate from one of *malice* on the part of bad actors to *recklessness* on the part of scientists. Safety meant that when it came to biological risk, the call was now coming from inside the house. And after a series of embarrassing mishaps, the US government in particular needed to recover its legitimacy by engaging in a new policy process.

The Deliberative Process

That new policy process would become known as the *U.S. Government Gain-of-Function Deliberative Process and Re-*

search Funding Pause on Selected Gain-of-Function Research Involving Influenza, MERS, and SARS Viruses, or the "deliberative process" for short. The process began with a funding pause on 21 experiments funded by the NIH on influenza and the—then—two coronaviruses believed to be of most concern for their pandemic potential, SARS and MERS. The coronavirus pause would be cut short almost immediately, with a selection of those experiments resuming within a month of the pause due to the perceived importance of developing and improving animal models for coronavirus research.

The deliberative process was designed as effectively a conversation between three groups, with input from the public (and recall what we mean by "public"—tickets to DC can be cheaper than those to Geneva, but still require free time and resources). First, two meetings were scheduled to be convened by the National Academies of Science, Engineering, and Medicine (NASEM), the first of which kicked off the deliberative process, and the second of which collected views and responses to the process later on. Second, the NSABB convened a series of meetings and commissioned a risk and benefit assessment of GOF research as well as an "ethics white paper" to establish an ethical framework around GOF research. And third and finally, the White House Office of Science and Technology Policy oversaw the process as the ultimate group to which the other two would report and issue recommendations.

This deliberative process took, all told, a little over two years, beginning in late 2014 and ending in early 2017; the conversion of the results of the process into policy by HHS would take almost another year, ending in December 2017 with the lift of the pause. The details of the risk assessment are voluminous—some 1,000 pages in length.[12] The principal author of that assessment has since told me that this is in fact a weakness of the assessment; he likened it somewhat cheekily to the Bible in the sense that "almost no one has read it in its entirety, and people just get what they want from it without reading the rest." Nonetheless, that assessment did its job, resulting in a draft report from the NSABB that would inform government policy.

That report found that while there are many kinds of studies that induce changes in transmissibility, virulence, and host range into viruses (i.e., GOF research), only a small number of them pose real threats to the world. It also revealed that existing policy did not cover all GOF research, and concluded that at least in theory, there are some kinds of GOF research that ought not to be funded or conducted. From this, the authors developed eight principles:

1. "The research proposal has been evaluated by a peer-review process and determined to be scientifically meritorious, with high impact on the research field(s) involved."

2. "The pathogen that is anticipated to be generated must be judged, based on scientific evidence, to be able to arise by natural processes."

3. "An assessment of the overall potential risks and benefits associated with the project determines that the potential risks as compared to the potential benefits to society are justified."

4. "There are no feasible, equally efficacious alternative methods to address the same scientific question in a manner that poses less risk than does the proposed approach."

5. "The investigator and institution proposing the research have the demonstrated capacity and commitment to conduct it safely and securely, and have the ability to respond rapidly and adequately to laboratory accidents and security breaches."

6. "The results of the research are anticipated to be broadly shared in compliance with applicable laws and regulations in order to realize their potential benefits to global health."

7. "The research will be supported through funding mechanisms that allow for appropriate management of risks and ongoing federal and institutional oversight of

all aspects of the research throughout the course of the project."

8. "The proposed research is ethically justifiable."[13]

These recommendations were submitted to the White House, along with the 1,000 pages of risk analysis, public comment, NASEM deliberations for two multiday meetings, ethics white paper, and associated scholarship.

The resulting policy framework issued was called the *Recommended Policy Guidance for Departmental Development of Review Mechanisms for Potential Pandemic Pathogen Care and Oversight*, or P3CO for short (if you say it out loud, there is an even chance you'll say the name of a *Star Wars* character). That policy framework arrived at a series of principles for how and when to fund GOF research:

1. The proposal or plan for such a project has been evaluated by an independent expert review process (whether internal or external) and has been determined to be scientifically sound;

2. The pathogen that is anticipated to be generated by the project must be reasonably judged to be a credible source of a potential future human pandemic;

3. An assessment of the overall potential risks and benefits associated with the project determines that

the potential risks as compared to the potential benefits to society are justified;

4. There are no feasible, equally efficacious alternative methods to address the same question in a manner that poses less risk than does the proposed approach;

5. The investigator and the institution where the project would be carried out have the demonstrated capacity and commitment to conduct it safely and securely, and have the ability to respond rapidly, mitigate potential risks and take corrective actions in response to laboratory accidents, lapses in protocol and procedures, and potential security breaches;

6. The project's results are anticipated to be responsibly communicated, in compliance with applicable laws, regulations, and policies, and any terms and conditions of funding, in order to realize their potential benefit;

7. The project will be supported through funding mechanisms that allow for appropriate management of risks and ongoing Federal and institutional oversight of all aspects of the research throughout the course of the project; and

8. The project is ethically justifiable. Non-maleficence, beneficence, justice, respect for persons, scientific

freedom, and responsible stewardship are among the ethical values that should be considered by a multidisciplinary review process making decisions about whether to fund research involving PPPs.[14]

Those principles were taken on by the agencies that fund GOF research so as to adapt them to the unique circumstances of each agency and its funding requirements. Only one agency ever released a final policy: HHS. This makes sense, as that agency is to date the overwhelming funder of GOF research not just in the United States but also worldwide. The agency's policy is to refer GOF research of concern to a department-level review, where a panel will review the proposed research under a series of principles wherein GOF research will be funded if the following is true:

1. The research has been evaluated by an independent expert review process (whether internal or external) and has been determined to be scientifically sound;

2. The pathogen that is anticipated to be created, transferred, or used by the research must be reasonably judged to be a credible source of a potential future human pandemic;

3. An assessment of the overall potential risks and benefits associated with the research determines that the potential risks as compared to the potential benefits to society are justified;

4. There are no feasible, equally efficacious alternative methods to address the same question in a manner that poses less risk than does the proposed approach;

5. The investigator and the institution where the research would be carried out have the demonstrated capacity and commitment to conduct it safely and securely, and have the ability to respond rapidly, mitigate potential risks and take corrective actions in response to laboratory accidents, lapses in protocol and procedures, and potential security breaches;

6. The research's results are anticipated to be responsibly communicated, in compliance with applicable laws, regulations, and policies, and any terms and conditions of funding, in order to realize their potential benefit;

7. The research will be supported through funding mechanisms that allow for appropriate management of risks and ongoing Federal and institutional oversight of all aspects of the research throughout the course of the research; and

8. The research is ethically justifiable. Non-maleficence, beneficence, justice, respect for persons, scientific freedom, and responsible stewardship are among the ethical values that should be considered by a multidisciplinary review process in making decisions about whether to fund research involving PPPs.[15]

If the department-level review by HHS looks like a carbon copy of the PC3O policy recommendations, which look like a carbon copy of the NSABB recommendations, which in turn look like a carbon copy of the 2013 H5N1 policy with the single addition that the work should be ethically justifiable, one might ask what everyone spent almost half a decade working on. It is also worth noting that where we began with a funder *and* a department-level review, that work was consolidated into a single department-level review involving one group rather than two. So the outcome appears to be one new principle—which, as I will describe in a future chapter, may or may not be necessary—and fewer steps of review for GOF research.

As I said at the start of this chapter, this is a sketch of the policy that applies to GOF and only GOF. There are many other policies that regulate the conduct of GOF, and so much of this research either never happens or is tightly controlled when and where it does. The policy that does describe GOF, however, has remained largely unchanged over the last decade. That is, the additional layer of screening that applies is one in which the department in charge determines if the research has scientific merit, studies a credible future pandemic, is the only way to perform such a study, has risks that outweigh its benefits, and follows existing regulation and ethics. This is what makes GOF approvable in the United States, which, as the largest funder of GOF research as defined by these kinds of policies, has defined much of the GOF landscape to date.

6

CURRENT CONTROVERSIES

The policy debate that ended in 2017 did not end the discussion about GOF research. The work continues, and in this chapter I'm going to go over some of the existing controversies that remain for this debate. In particular, I will describe and review the newest policy released by the US government on GOF research, which brings us back to the convergence between GOF and dual use, and which was published on May 6, 2024—while I was waiting for editorial to return the manuscript of this book. (I'd like to thank the White House personally for not publishing this during the typesetting stage, which would have been exceedingly chaotic!)

The NSABB began meeting again in 2022 and commissioned a report that would be released in March 2023 outlining new potential options for the US government.[1] The board offered a sweeping set of 11 recommendations, which I'll describe briefly here. First, it recommended some harmonization, particularly of the P3CO policy with other dual-use research policies, removing the exceptionalism of the former by incorporating it into the latter. The second was that while the P3CO was still narrowly tailored to existing PPPs, the NSABB recommended the policy be expanded to include any research that would risk creating a pathogen that could spread uncontrollably (an ePPP), cause significant harm to humans, and pose a risk to public health systems or national security.

The third and most controversial for scientists was its recommendation to remove the P3CO policy exclusions for vaccine and surveillance research. The board found that these might be considered blanket exclusions, which as discussed previously, risk making the policy so narrow as to cover nothing. Removing these effectively would eliminate an easy way for a study to escape review, and require it to undergo the same risk and benefit assessment as everything else.

The fourth recommendation articulated a set of criteria for streamlining the process and clarifying the roles. The fifth sought to clarify the uniqueness and risk reduction clauses of P3CO, and cast them in terms of promoting benefits and eliminating risks, allegedly in harmony with the Belmont Report, which guided the creation of US human research ethics in the 1970s and 1980s. It also outlined a list of documentation the US government should produce on standard operating procedures, and educational materials, among others, to guide investigators and regulators in assessing the risks and benefits of GOF research.

The sixth recommended additional transparency, following the contours of a current controversy I'll describe below. The seventh sought to align the P3CO policy with National Security Memorandum 16, which focuses on food and agriculture. This is a long-standing issue in the NSABB's work, which has until now almost exclusively revolved around direct, human-transmissible diseases and not the security and safety risks posed by research on animals and plants. The eighth reiterated the need to continue sharing experiences in regulating dual-use research and engage stakeholders (remember who the "public" is here) as policy develops.

Recommendation 9 would remove the phrase "directly misapplied" from the US government's current definition of dual-use research of concern. While I've largely skipped any additional movements on non-GOF dual-use research, this would remove a limit on consideration of the dangers of scientific research, expanding it beyond only those that result from the use of the scientific discovery directly. Recall that the polio experiment was less concerned with polio than it was with future developments; under the current regime, that wouldn't be considered research that would be subject to review, but if this change occurred it would allow the US government to consider the broader impacts of the work.

Recommendation 10 would expand the scope of review for dual use to any pathogen, not just those on the US "select agent list." Recommendations 11 and 12 closed out the report with, respectively, a call to continue to engage professional groups on adopting their own dual-use policies and to ensure that dual-use research funded by the US government—including GOF research—is conducted in the United States where possible and subject to extra scrutiny when done abroad.

These current developments are all situated in the aftermath of the P3CO policy and concern the future directions of not only GOF research but also dual-use research more generally. Despite a decade of debate, some questions about these kinds of research remain difficult to answer or involve tendentious solutions. The three I'm going to address here are those that I think remain the most important. The first is a substantive question—that is, one that bears on the actual research itself. That question is, What is the true value of GOF research? As I discussed in a previous chapter, ferrets are notoriously difficult animals to study, and what we learn about flu viruses is attenuated partly by our troubles with ferrets. Research with other pathogens face different issues, and so rather than focus on influenza, I am going to zoom out to larger questions about how and why we choose GOF research, and what creating a novel pathogen tells us about the risks of pathogens in the wild.

I'll then follow up by addressing one of the chief criticisms of the P3CO policy as it was implemented, which is the issue of transparency in the way departments and HHS in particular review GOF research. Even for those satisfied with the conclusion of the deliberative process—and they are rare indeed—the *conversion of the process into policy* has alarmed some observers. Most of the review panel's dynamics are not open to the public either in its roster, procedure, or deliberations and final decisions.

Finally, there are continued questions from within the scientific community about how any policy can be implemented without overreach. In particular, concerns have recently been raised around the idea that GOF research is "reasonably anticipated" to create a novel virus. In this final section, I'll show that this alleged vagueness serves an important interpretative role, but also functions to shield scientists from restriction as much as it threatens them. Moreover, it highlights a future direction for the regulation of science, which is to establish bodies of regulatory evidence so that decision-making can be assessed on its merits rather than appear to be arbitrary as it so often does to outsiders.

The True Benefits of GOF

Why do GOF research? This is a question that is surprisingly still difficult to answer. Let's start again with the basics. Proponents highlight three central benefits of GOF research:

1. Discovering a new variant of a potential pandemic pathogen raises the alarm on the possibility of an emerging pandemic

2. GOF research allows us to determine mutations to look out for in disease surveillance

3. Identifying pandemic diseases before they emerge assists us in developing countermeasures such as medicines and vaccines.[2]

Over the last few chapters and elsewhere, I have discarded the first of these already.[3] It's not clear that even prior to COVID, the world needed more alarms raised. If action wasn't taken, it wasn't because of the endless news articles that read "the world is not ready for the next pandemic," the millions spent on biosecurity research and preparedness, the UN meetings and initiatives, the US Global Health Security agenda, and the near misses of SARS, H5N1, and 2009 H1N1.[4] The global order's perspective on public health is an example of what is often termed "panic and neglect," and something *else* beyond the publication of a paper in *Science* or *Nature* is required.

There is a term in the social sciences that can do some explaining here: "political imagination."[5] Publications in high-impact journals or mainstream outlets (as appropriate for the class of person that a high-profile scientist is) such as the *New York Times* are a currency to researchers. They are good for their careers and the institutions they serve—universities, yes, but also funding agencies. But this isn't merely a cynical choice on the part of scientists: the kind of currency the academy selects for individuals who are particularly adept at creating the kind of value that is prized by the institutions they ultimately come to rest in. I'm thus not saying that Fouchier or Kawaoka could do differently, but secretly choose not to. I'm saying that the kinds of people Fouchier and Kawaoka are largely precludes them from imagining anything else. The same can be said about Ab Osterhaus, one of Fouchier's mentors; Susan Wolf at the University of Minnesota, the first bioethicist ever nominated to the NSABB; Michael Osterholm, a vocal critic of GOF research from the University of Minnesota's Center for Infectious Disease Research and Policy; the University of North Carolina at Chapel Hill's Ralph Baric, who came under fire for his coronavirus research and its connections to various theories about the origins of COVID; or Susan Weiss, a brilliant coronavirus researcher who does *not* conduct GOF research, but nonetheless has been subsumed into discussions of COVID and biosecurity. These luminaries are selected, in a close analogy to evolutionary pressures in nature, precisely because of the kinds of action they produce.

GOF research is therefore not only incapable of changing minds on pandemic preparedness but unable to do so *by design* as well. The catch is that the assertion that it can do so will likely remain, despite evidence to the contrary. This is not a matter of personal choice in isolation but rather a structural feature of what scientific research is, and how it interfaces with our politics. Concerns about the role of basic biodefense research have persisted for decades, yet largely from outside the elite circles of the scientific world, and absent some greater change I suspect this will remain the case.[6]

Let's now turn to the second justification for GOF. Recall just what it is that GOF research is supposed to provide us with knowledge about. We are attempting, essentially, to predict the evolutionary future because doing so is instrumentally valuable to preventing a catastrophic disease pandemic. What kinds of mutations might plausibly arise within a given viral lineage, and how might those mutations lead to changes in the viruses' ability to infect and transmit between humans? Any critique of this research must, to be sure, take this process seriously.

Yet a look at accounts of purported successes of GOF research raise serious questions about its ability to support the kind of GOF experiments we're worried about at the level of policy, much less those that would be subject to regulation or restriction on funding. Let's examine closely a case given in the literature to see why this is. In

GOF research is therefore not only incapable of changing minds on pandemic preparedness but unable to do so *by design* as well.

an editorial published at the beginning of the deliberative process, a consortium of researchers led by Stacy Schultz-Cherry, a virologist at the WHO Collaborating Center for Studies on the Ecology of Influenza in Animals at St. Jude's Children's Research Hospital in Memphis, Tennessee, defended GOF research for its capacity to inform decisions about which candidate vaccine viruses—samples of virus created and stored in preparation for development into vaccines—to develop for defense against seasonal and potential pandemic influenza strains.[7] The researchers give an example of a 2013 decision to develop a candidate vaccine virus from A/Cambodia/X0810301/2013, a flu virus that began circulating in the early 2000s and generated concern as the number of cases grew from 11 in 2005 to 26 in 2013. The team claimed that this decision was made based on the detection of certain genetic markers identifying the possibility that this virus would become a pandemic strain. Those mutations were

1. S133A and S155N, which had been found to have caused binding to mammalian sialic acid receptors; recall from the Imai paper that these are the receptors that flu needs to bind to in order to get into mammalian cells

2. S123P, which does the same as S113A and S155N in conjunction with other mutations

3. K189R and Q222L, which do the same again

4. N220K with Q222L, which has been found to increase droplet transmission in ferrets

The researchers asserted that the WHO Collaborating Center chose its candidate vaccine virus from these and that this could not have occurred without GOF research.

But a closer look at the evidence they supply paints a slightly different picture. Starting with S133A, S155N, and S123P above, the researchers cite three papers to support their conclusion. The first of these is an experiment published in 2010 on a live, attenuated HPAI H5N1 strain. This is a strain of H5N1 that has, in another application of the serial passaging that made the original GOF papers the object of controversy, been depleted of its capacity to cause significant disease for the purpose of safe research and development. In fact, the virus in question was another vaccine virus, and these are typically wild viruses that are attenuated so they are ready for vaccine development.

The second paper was conducted by Kawaoka's team at the University of Wisconsin at Madison in 2006, and is much closer to the canonical GOF research of 2011. There is a twist here, however. Following similar methods as described in the previous chapters, Kawaoka's team generated a library of eight HA H5N1 viruses to test in vitro; no ferrets were used in this study. Yet rather than creating a library of possible evolutions of HA H5N1 like they did in 2011, the library was built from eight viral samples that were isolated from humans. That is, the researchers were not creating novel viruses to test but instead testing the properties of the hemagglutinin arms of viruses (the bit that allows flu to enter human cells) that were already able to infect humans, combined with the neuraminidase of a Vietnamese lineage of H5N1, on top of the rest of our old friend A/Puerto Rico/8/34. This paper, then, is not in the business of creating novel viruses except insofar as the researchers are using reassortant viruses to make sure that the *only bits of flu they are testing* are in fact the hemagglutinin arms. They aren't trying to create, to come back to our definition, novel viruses with additional properties; they are trying to figure out the properties that existing viruses already have.

The third and final paper involves the attempt to distinguish, again, between different versions of hemagglutinin to determine which mutations make the virus particularly suitable for infecting mammals. This study, published by researchers at the Vaccine Research Center of the National Institutes of Allergy and Infectious Disease in 2007, sought to determine which variants of HA were most likely to infect mice. To do so, though, the researchers used a method whereby a *vaccine* strain of virus was manufactured to include the relevant genetic information from HA. So rather than the way we understand GOF research as a policy term, this research is not testing a potential pandemic virus; it is testing a vaccine strain of influenza that encodes a particular HA arm. So a further GOF experiment is implied, but unlike the Herfst or Imai papers, is not actually done; it's a version of the first half of those papers, but they stop before the researchers try and design a new live virus that causes disease.

I think the most likely explanation for this discrepancy is as follows. The problem seems to be that—to return to the start of this story—misunderstandings about policy definitions and scientific terms for "gain of function" are baked into this debate. I challenged the idea that GOF really is as ubiquitous in virology as some have made out, at least in the descriptions provided in the professional literature. If one has decided that GOF is expansive, however, as the virology community has done, then that third and final paper might be a vaccine strain of virus "gaining" a function in encoding an HA arm. Again, that's incredibly broad, but plausible given the stakes of the debate around what is and is not GOF research.

And this is where the oral history of the deliberative process comes in. The initial ask that signaled the beginning of popular support for a formal risk assessment on GOF was published by a group of scholars that referred to itself as the Cambridge Working Group.[8] I was one member of that group and drafted the first language of its consensus statement prior to a meeting in 2014. But that statement, which asked for a moratorium pending a deliberative process, was a mild request compared to a number of extreme demands made by critics of GOF. Those demands have included a moratorium on all pathogen research; reducing the number of high containment laboratories (including BSL-3 labs, the second-highest level of containment) to one per state in the United States at most; and even civil or criminal charges for scientists associated with a lab escape. These were actively discussed on the margins of the first workshop in the deliberative process held at the NASEM as well as at NSABB meetings in years to come. One might think that defending all virology as GOF is an adaptive strategy that pits the relatively narrow policy process that resulted in P3CO against the stakes of pathogen research writ large being curtailed.

But even with this in mind, the true value of GOF has still not really materialized. The GOF benefit analysis conducted during the deliberative process demonstrated that only in 9 out of 24 scientific and public health goals addressed for influenza was GOF/PPP uniquely useful; this number was only 3 in 13 for coronavirus (SARS and MERS) research.[9] So whether GOF research in practice answers any important question, much less anything unique, remains an unsolved issue. And this is going to continue to be an issue, as I'll discuss at the end of this book, as eyes turn to GOF once more.

The 2024 US Policy: GOF Is Dead, Long Live GOF

In May 2024, the White House issued a new policy that covered GOF research. This was presented as a step in unifying the issues of GOF with those of dual use, which, as discussed previously, was a progenitor to the current debate. The policy replaces the P3CO policy and a number of other dual-use policies on the books. This shouldn't be understood as progress toward some kind of rule or law, however; these guidelines are, as I noted earlier, a more specific kind of policy that applies to review of federal grants at the level of individual departments.

What makes this policy fascinating as the climax of this story is that the term "GOF" does not appear anywhere in the document. How can a policy that acts as a direct replacement of *the* GOF policy, the product of more than two years of debate, not mention this term at all? The answer lies in the architecture of the policy and its integration with other areas of biosecurity and biosafety. The move that has been made is away from a *kind of research*, to the *outcome* of scientific experiments.

The new policy lists two categories of experiment that are subject to review. The first, "Category 1 Research," concerns experiments that are now considered dual-use research and represents an evolution in dual-use policies from almost a decade ago. This is interesting in itself, as agricultural and veterinary threats have been elevated in their importance under the policy. But this isn't really GOF: to be subject to category 1 research review you have to be doing research with a particular kind of pathogen toward a particular kind of experimental outcome.

Enter "Category 2 Research." What counts as category 2 is research that

1. "involves, or is reasonably anticipated to result in, a PPP";

2. "is reasonably anticipated to result in, or does result in, one or more of the experimental outcomes or actions" the policy is concerned with; and

3. "is reasonably anticipated to result in the development, use, or transfer of a PEPP or an eradicated or extinct PPP that may pose a significant threat to public health, the capacity of health systems to function, or national security."[10]

Some unpacking is needed here. That first requirement appears relatively self-explanatory—*any* pathogen that will be modified such that it is reasonably anticipated to result in a PPP is covered by the policy. The second part concerns four specific kinds of experiments, namely those that are expected to:

1. Enhance transmissibility of the pathogen in humans;

2. Enhance the virulence of the pathogen in humans;

3. Enhance the immune evasion of the pathogen in humans such as by modifying the pathogen to disrupt the effectiveness of pre-existing immunity via immunization or natural infection; or

4. Generate, use, reconstitute, or transfer an eradicated or extinct PPP, or a previously identified PEPP.[11]

This leads us to: What is a PEPP? I mentioned this very briefly earlier, as a new term representing "pathogens of enhanced pandemic potential," which are PPPs that result from experiments that enhance the transmissibility, virulence, or immunity evasion of a pathogen. That is, PEPPs are PPPs caused by GOF or GOF-like research. So under the third criteria above, category 2 research is concerned with any experimental or scientific technique that generates PEPPs; put another way, anything like GOF that creates a PPP from another pathogen.

So GOF is gone. What is left are PEPPs, which are viruses created from research *like* GOF.

What happens next follows in very similar ways to previous policies. There is a department-level review policy within federal funding agencies to ensure that category 2 research is justified. The policy—responding to controversies about transparency in GOF research—provides a schematic for how agencies should structure review committees, and then directs those committees to review category 2 research with a series of principles that are more or less identical to P3CO.

This isn't to say that there aren't promising developments, however. For one, the policy directs agencies to, where possible, close loopholes and enforce policy even when research isn't being conducted using federal funds. This responds, I suspect, to a false-positive GOF study that emerged from the New England Infectious Disease Laboratory that some claimed was GOF research but did not create a virus of enhanced virulence or transmissibility.[12] But it also reflects an extended trend in research ethics of attempting to make US research regulations cover more than merely federal rules for funding requirements, as most developed nations have their research regulations enshrined in laws that cover *any* research carried out in their jurisdiction.

Moreover, the policy responds to two outstanding concerns around transparency and reasonableness that became bugbears for the P3CO policy after its release. This new policy attempts to clarify those concerns to reflect some of the ongoing tensions around GOF research, but in a way that at least nods to some of the government's critics.

Transparency and Integrity

Much of the new policy emerges from criticisms of the P3CO policy and how the review works in practice. That is, while P3CO contained a schematic for when funding decisions about pathogen research should receive additional review and some note as to how this process should resolve, the identity and composition of the groups that made these decisions along with the kind of deliberations they had was closed. So closed were these decisions, in fact, that there was no public account of how many studies had been reviewed, how many were approved, and why some (if any) were rejected.

This was viewed as constituting a lack of transparency to a policy that, for many, was a hard and bitter fight to win. Instead, what was implemented was a commitment to some kind of decision-making process around safety and security in funding GOF research, but no account of whether the framework was being implemented in any meaningful way—and by whom.

The response from HHS officials around this problem was to note several things. First was that at least from a security perspective, there was a paradox around the open assessment of these kinds of research protocols. That is, if there were important security issues to discuss, the typical move would be to not report those issues but rather keep them confidential.

The second response was to point out that, by and large, review committees are not open affairs. For example, internal review boards (IRBs) rarely openly report their deliberations or decisions. Nor do they name their committee members. In general, this is because the kinds of information being reviewed by IRBs are protocols for research, and investigators have an interest in not openly reporting what they are planning to do. Especially for cutting-edge research, there is a general presumption that information of this kind will be private unless it needs to be registered in compliance with funding obligations, such as on clinicaltrials.gov. And these are typically for reasons of robustness in scientific research, not to ensure public review of protocols or IRB decisions. Whether IRBs should openly report deliberations is another conversation, but certainly as far as P3CO went, the argument was that while we presume all other reviews stay closed, this one would too.

Finally, concerns had been raised that revealing the information of the committee and its deliberation might have resulted in its being targeted for harassment or worse. It wasn't clear, at least at the time, why this might be. But the more recent events around GOF, as the debate about GOF has become tied to more colorful theories surrounding the origins of COVID-19, has seen the harassment of scientists online and in person.[13]

The new policy attempts to obviate some concerns while remaining firm on other responses by describing the process *procedurally*. That is, while the new policy does not give the names of committee members, nor require this to be done, it gives information as to the makeup of review committees. Review must be conducted by committees of officials who: don't report to the heads of agencies who fund category 2 (the category formerly known as GOF) research; have scientific, biosafety, biosecurity, medical countermeasure, law enforcement and national security, ethics, public health, biodefense, select agent, and other areas represented; and give other agency and government officials input into the process. That is, much like the makeup of IRBs, *who is represented* is known, but the representatives themselves are not.

Such a move attempts to sidestep the concern that if no one can see the review committee's deliberations, the committee might not be doing anything of note—or worse, it might be applying an inappropriate standard to the studies it reviews. It does so, moreover, in a way that recognizes—I think correctly—that the identity of those members is not the determining factor in the problem of inappropriate review. And indeed, as we'll talk about next, this policy is, like P3CO, subject to a number of ambiguities that, outside of establishing an open account of the committee's deliberation, might be used to over- or under-regulate GOF research.

Reasonable Anticipation

A central concern for critics of the P3CO governance was its ambiguity. In particular, there was one term used in the policy and the department-level review that came after at

HHS that concerned some members of the scientific community: "reasonably anticipated."

This phrase actually appears a number of times in the guidance on GOF research over the last decade, but it especially concerned people in the context of what was covered by the P3CO review process. That is, the guidance states that "proposed intramural and extramural life sciences research that is being considered for funding and that has been determined by the funding agency as *reasonably anticipated* to create, transfer, or use enhanced PPPs is subject to additional HHS department-level review as outlined herein."[14] That use of "reasonably anticipated," emphasized by me in the quote above, was of concern to scientists because there is no articulation of what might be covered as "reasonable." The fear, then, was that decision-makers reviewing GOF research might take "reasonable" as a reason to overregulate and suppress scientific research to the detriment of the field and humanity.

In response, the new policy attempts to clarify specifically what it means by reasonably anticipated:

> *"Reasonably anticipated"* describes an assessment of an outcome such that, generally, individuals with scientific expertise relevant to the research in question would expect this outcome to occur with a non-trivial likelihood. It does not require high

confidence that the outcome will definitely occur but excludes experiments in which experts would anticipate the outcome to be technically possible, but highly unlikely.[15]

The implementation guide paired with the policy notes that this definition does not require more than 50 percent confidence; what it does say, in keeping with what I have described throughout this book, is that scientific experiments typically frame hypotheses that set *expectations* about what an experiment is meant to produce. That is, the Herfst paper *expected* to create a mammalian-transmissible, highly pathogenic H5N1 virus. Moreover, this was not a 50–50 chance of doing so but rather a *reasonable* chance in the sense that most rational people can agree that if someone claims they intend to do something in science they are qualified and skilled enough to do, we should treat them as if they will succeed.

I suspect, however, that this will not sate critics of GOF, or category 2 research policy. That is, what they appear to be asking for is a strict threshold, say, of risk of release, or of an expected number of casualties if a release happened, against which to make assessments. This is speculation on my part concerning the *unit* of the measure they are asking for, but I think it is, if you will, reasonable to suspect that a "scientific"—in the sense of *objective, impartial*—measure of risk is what is being sought.

There's a lot of reasons to be concerned about government overreach, especially in science. The US government in particular has a long and complicated history that—despite some appeals to the unrestricted nature of "basic research"—often finds controversial research in the line of fire. This includes the suppression of environmental research conducted by the US Department of Agriculture during the Bush administration, laws that prevent certain kinds of stem cell research, and laws forbidding the use of federal funds to explore firearm safety and deaths.[16] So on the face of things, there are certainly many cases we can examine of the overregulation of science.

But here, proponents of GOF have frequently protested too much. One example of this came on the eve of the first NASEM meeting in 2014 that marked the start of the deliberative process. A survey was published in *mBio* alleging to report the attitudes of students about their career prospects in view of the pause on funding for GOF research. That survey, drawn from a convenience sample on social media and within the author's institution, found that of the 156 respondents, 28 percent said they were less likely to work on influenza, SARS, or MERS as a result of the GOF debate.[17]

Yet as even the author at the time noted—while claiming that these kinds of chilling effects should be factored into the risk-benefit calculus for GOF—in addition to being an ad hoc survey conducted by a researcher with no survey experience or expertise, the overall validity of these results seems to be questionable. The "debate" at the time covered 18 projects, though by the end of December 2014 that number would be considerably lower, as coronavirus research was largely exempted from the pause.

Even setting aside methodological problems, what might explain this particular response by students? One easy answer is framing effects. The virology community has largely framed GOF regulation as overreach, and so we can imagine that its students likewise will frame that regulation as overreach. Importantly, it does not matter whether the regulations *are* overreaching, merely that they are perceived to be.

So too with "reasonably anticipated." The framing of that phrase given by virologists is that GOF research might lead to overregulation. But why should we anticipate that particular direction of governance in the presence of ambiguity? Given that the development of GOF research has been a decade in the making, and has moved almost nowhere in that time, it seems like the opposite might be equally, if not more, likely. That is, we might reasonably anticipate that "reasonably anticipated" is not going to lead to overregulation but rather underregulation given that the regulation continues to happen in-house at the department level in the US government, is not terribly transparent, and relies on a framework that it is quite permissive otherwise.

And permissive it is. While considerable worry has emerged about "reasonably anticipated," much less has emerged over the fourth principle in the P3CO framework—and the new policy—which states that GOF research may be conducted if there are no feasible, equally efficacious alternative methods to address the same question in a manner that poses less risk than does the proposed approach. In a strict sense, though, there are *no* methods that answer *exactly* the same questions as GOF. So in principle, then, we might worry that all GOF research, so long as it isn't projected to carry risks that are outlandish compared to its expected benefits, might pass muster.

A key issue here is less that there is a lack of scientific grounding for these terms and more that there is no history on which to develop these kinds of normative decisions. There may not be a definitive, scientific grounding for expectations about the chances that a study will lead to a PEPP ahead of time—*at all*. It's possible that the only way to assess this is on a case-by-case basis, building up what is effectively a volume of "case law" about GOF governance. This body of work would include an assessment of research that is intended to create enhanced PPPs and appears to succeed (such as the Herfst and Imai papers); research that intends to create enhanced PPPs and fails; and research that does not intend to create PPPs but does so anyway.

A key issue here is less that there is a lack of scientific grounding for these terms and more that there is no history on which to develop these kinds of normative decisions.

Expecting scientific grounding in science policy can be understood as a neat way to crash a policy process altogether. It is, in a way, a trick we see in other areas like tobacco health research, environmental research, and climate change forecasting to prevent meaningful regulation from taking place and thus developing into robust and thoughtful decisions. That is, uncertainty is painted as a reason to forgo meaningful regulation or reform. And while we shouldn't assume that virologists are arguing in bad faith, it is important to recognize that when it comes to ethical and policy decisions as they appear in GOF research, asking for certainty in the face of risk should be taken with a decent pinch of salt. In regulatory debates there are always fears of overregulation, but the historical record on GOF shows that, by and large, these worries have never come to pass—and, if anything, have tended in the opposite direction.

7

THE ETHICS OF GAIN OF FUNCTION

At the end of the last chapter I referred to GOF as an ethical decision. In this chapter, I'm going to talk a little bit more about what that means. It is particularly relevant to the 2014 debate in which an ethics white paper was commissioned by the NSABB in addition to its risk and benefit assessment.[1]

A common question by scientists, policymakers, and the public alike is, "Why an ethics of GOF?" The basic answer is that GOF research implicates a number of values, and ethics is the study of values. These values include well-being (or utility) in the form of human and animal welfare, and the benefits of scientific research; equality in the form of the distribution of benefits and risks imposed on different people; liberty in the form of the freedom of scientists to conduct their research projects; and security in the assurances we can have that certain kinds of catastrophes won't befall us. These values are not monolithic and can conflict with each other.

Ethics, as one definition will have it, can be descriptive in nature; we find out what values people have, and why. But it doesn't just tell people what their values are; it can tell us what our values *should be*, and how we should act accordingly. It can help us reason through difficult or new decisions and arrive at a decision on how to act.

There are a number of ethical questions raised by GOF research. The first is how we balance the different values at stake, the potential harms of a lab release or the misuse of scientific research and the freedom of scientists and potential benefits of the research. The second is about whether the way those benefits and harms are *distributed* matters for GOF research. The third and final is if there's anything particularly special about the kind of problem GOF research is that requires us to make calculations about extremely rare and unlikely but potentially catastrophic kinds of risks. We'll tackle each in turn here.

Plural and Conflicting Values

As a preliminary beginning, it might look on its face like one way to understand the dilemma of GOF research is to simply add up the number of people we expect to be killed if a release of a novel virus created using GOF research

GOF research implicates
a number of values,
and ethics is the study
of values.

were to occur, and weigh it against the potential benefit—for example, the number of people saved if the research lead to a new vaccine—we stand to gain if we pursue the research. It's a simple equation in theory: lives minus deaths. Of course, things are never that easy.

First, measuring the chance of a lab release is quite difficult. This is because these are rare events; the number of lab releases in the world, as a proportion of the number of labs and the number of people working in them, is extremely small. The number of releases that have lead to an outbreak is smaller again, and the number of outbreaks that have gone global is smaller still. In fact, with the exception of the theory about COVID-19 we'll discuss in the next chapter, the only example given in the GOF debate is the 1977 influenza pandemic, which is believed to have been a potential lab release.[2] The risk assessment portion of the deliberative process came up with a rough rate, in the ballpark of 1 in 54,000 for 10 labs doing GOF research for 10 years.[3]

Second, even if you did estimate this risk, the benefits of GOF research are even harder to assess in a way that makes them easy to compare against the risk of a lab release. I'll discuss distribution problems in the next section, but even just understanding what these benefits look like and how many people they'd save over a particular time period—say, 10 years—is exceedingly difficult. That's because science doesn't exist in a vacuum and depend on a single event, like a lab release. You would have to estimate the chance in a 10-year period that the right experiment happened, for the right pathogen, at the right time, and that the rest of the science and technology landscape was aligned.

This doesn't mean these benefits aren't real—that problem exists, as we've explored, but this is somewhat different. It means that *quantifying* those benefits as numbers that we can add or subtract from the risk of a lab release is extremely difficult to do. So difficult, in fact, that the estimate of the benefits of GOF research in the deliberative process was a qualitative rather than quantitative process: researchers assessed case studies and interviewed experts to determine how these benefits might emerge.

But a third problem is that the kinds of values actually in play aren't easily comparable among each other, and go beyond risks and benefits. In the case of GOF, this comes up most often in the case of scientific freedom. In the twentieth and twenty-first century especially, a central norm for science has been that scientists should be free to conduct and communicate research of their choice. While this freedom isn't unlimited—for example, conducting unethical research on human subjects—it is frequently taken to be permissive in the absence of a particularly weighty reason to do otherwise.[4]

How serious does a threat need to be, then, in order to justify overriding scientific freedom? One way to conceive of this might be when there is a strong national interest in preventing some deeply harmful event from occurring. In this case, critics of GOF are typically worried about an accidental or deliberate release of a recombinant pathogen that could seed a global pandemic.

An implication of this, given our recent experiences with COVID-19, is that the threat of an infectious disease to a nation's interests and structure need not be an existential or even catastrophic risk to be serious enough to act. A coronavirus, after all, with a mortality rate around 1 percent successfully immobilizing a nation was a strong contributor to rising radicalism and challenges to the state, and will inflict long-term harms on the globe.[5] As I will look at in the next section, the origins of COVID-19 are almost certainly a wild animal host. But what COVID-19 does give us is a biological benchmark, of a kind, for what an infectious disease outbreak can do to national and international integrity in the twenty-first century. The damage caused by the 1918 pandemic gives a similar, historic reference, and was the basis for comparing potential GOF impacts in the risk and benefit analysis in the deliberative process.

This is not to say, however, that we ought to censor scientists who have done GOF research. There are a range of reasons not to do so, but the biggest one is the distributed nature of contemporary scientific research. That is, once the research is done, the cat is out of the bag. This shouldn't mean that a deeply scary, world-ending experiment that emerged unexpectedly shouldn't be censored, but as a routine policy, it is an expensive process that carries serious risks of abuse and can often be replaced with other, less invasive means.

Those means are likely at the level of funders. There is potential benefit from GOF, but it is frequently difficult to understand and grounded in ambiguities around methods and our fundamental knowledge of viral genetics. Reprioritizing experiments and funding more basic understandings of ferrets as well as epistatic interactions between genes in flu would fulfill this kind of obligation.

How does this sidestep the problem of scientific freedom? Scientists should largely be free to conduct and communicate research of their choice but, at least as is commonly understood, they don't have a right to be *funded*. Plenty of scientific fields are underfunded, and their practitioners rely on cheap methods or institutional support to conduct their research. The R01 grant program, the flagship of the NIH, has a success rate of roughly 20 percent. Yet we don't think that the four out of five scientists who aren't funded have had their rights violated. And many of those failed applicants aren't doing bad science but rather science that doesn't fulfill the priorities of a funder or the preferences of the reviewers.

Where does the current GOF policy stand? As a funding mechanism, it seems to be applied at close to the right level of governance. That is, it doesn't violate scientific freedom but instead serves as an additional level of oversight, and in a form we largely understand and accept. Science policy does already involve decisions grounded in concerns of safety, and so this is consistent with other forms of governance.

In terms of its ability to navigate those risks and benefits of GOF, however, it isn't clear that the new policy is appropriately aligned. This is in part because there are some criteria that appear to rule out the vast majority of work from consideration according to the policy so long as the research targets a unique scientific question. But the bigger concern some have is that its prescriptions for weighing risks and benefits are ambiguous, and the process by which reviewers arrive at their judgment isn't transparent.

The Public Good of Science

A second concern for the ethics of GOF focuses on the winners and losers of scientific advance. Think again about GOF research and what the scientists who conduct this research claim is its purpose. The "win" according to proponents is increased disease surveillance, vaccines, and medical countermeasures. The "loss" according to critics is a global pandemic.

But note that the "win" as described doesn't really account for *who* wins. And here we have a recent case to ground us: vaccine nationalism during COVID-19. By "vaccine nationalism," I mean the practices of rich nations during the pandemic of buying up and hoarding vaccines, preventing other nations, often in emerging hot spots, from accessing them. Some countries, such as Canada, bought more vaccines than they had people, multiple times over. At times, such as in the United States, the inappropriate distribution of vaccines within a country led to their destruction and loss—vaccines that might otherwise have gone to low-income countries if better decisions about global allocation had been made ahead of time.

This kind of problem was also the subject of a debate in 2006 over the appropriate use of influenza A H5N1 samples. Indonesia, on the receiving end of an outbreak of H5N1, refused in December 2006 to share samples of the virus with international researchers. The scientific community was largely up in arms about the possibility that a nation would curtail the study of a pandemic virus, but Indonesia's rationale speaks directly to the issues raised by GOF research: the government of that nation maintained that it was providing samples to the largely Western community of scientists and medical countermeasure developers but was never the intended recipient of any innovation or discovery that emerged from those samples. The result of that debate was the Pandemic Influenza Preparedness Framework.[6]

While that framework has its strengths, however, it lacks the teeth to truly overcome the realities of innovation and technology in the twenty-first century. Legal analysis of the framework notes that the mechanisms of the treatment don't support a sufficient supply of vaccines to the WHO or its member states, nor the broad acquisition of technologies by states to ensure they can build a supply of vaccines against pandemic influenza. It also applies only to flu viruses, which would leave anything from the SARS and COVID families out in the cold—among the wide variety of bacteria, viruses, and fungi that might threaten us.[7]

The public good of GOF, then, from an ethical standpoint, is somewhat of a tricky problem beyond our difficulties in measurement. Advocates of GOF research claim that while a lab accident leading to a global pandemic is *possible*, it is hardly *probable* as critics might claim. Yet, at the same time, they articulate the potential benefits of GOF as *likely*, notwithstanding the difficulties faced. This is a kind of moral asymmetry around GOF that requires a bit of explanation.

But what about the probability that the governments of the world will come together to agree on a mutually beneficial vaccine manufacture and distribution network for pandemic disease, finance it, and maintain it against emerging threats? On the one hand, the answer seems to be "zero, or close enough to it." But this is itself too easy.

These aren't questions necessarily of probability but rather questions of human agency and interaction. There are probably political moments where the chances are much greater than zero, and times where they really are zero. It depends on a lot more than the stochastic nature of lab releases in modern science.

But another, and I think just as compelling, answer is a kind of technological optimism. We often assume the benefits of technology—not just in the United States, but all over the world. Eventually, the line of reasoning goes, someone will build the technology the right way, and it'll make it into our lives. It may not be fast, and it may never be complete, but it will occur. And that "will" does a lot of work ethically, for a lot of people. If we forgo research, then, the "will" becomes a "will not" and the supposed eventual benefits of science will never arise. This is a kind of global, technological version of the Gretzky quote "you miss 100 percent of the shots you don't take," but with pandemic diseases.

But that optimism obscures the fact that infectious disease research, even well intentioned, occurs in a global health framework that is deeply flawed. On the chance that one of my readers simply doesn't care about other people in other countries, note that even our national public health systems, I suspect, are inadequate to the task. Within the US, COVID-19 had its well-defined groups of winners and losers, and those groups mapped well-worn paths of injustice that are a core feature of American life. GOF research can't solve this, but we should adjust our expectations for GOF's potential to improve our lives—especially because those same weaknesses make it all the more likely that a laboratory release will lead to catastrophe.

Catastrophic Risk in a Globalized World

The ethics of GOF, finally, is an exercise in attempting to avoid a "catastrophic risk." A global pandemic of a virus with lethality of 10 percent, or 20 percent, would be a catastrophe of monumental proportions. It would alter the trajectory and development of our world for a very long time. But the chance that this will happen is exceedingly small. And our intuitions as moral agents tend to diverge, often strongly, from each other when it comes to probabilities so small and consequences so big. One way to imagine this is as a deeply unpredictable risk: like rolling a million-sided die where, if you roll anything but a 1, you get a small but nontrivial reward (let's say, $100); but if you roll a 1, everyone you know dies.[8]

The common way to view this problem would be to use *expected value theory*, which states that we act rationally when we maximize our expected value. Let's say you have 999,999 possible outcomes where you get $100, and 1 where everyone you know dies. As long as 1/1,000,000 multiplied by the value of your loved ones is less than $100, you ought to roll the dice. You can then solve this kind of problem iteratively to understand how many times you ought to roll the dice before the stakes get too high.

The *obvious* problem is there's no straightforward way to compare that money to the value of *your* loved ones for *their* sake—and so plural and conflicting values strike again! The National Highway and Transit Safety Authority has a formula for the cost of a human life in dollars, but that's not quite the same. There might be a social value to your family we could quantify in terms of work hours or other contributions to society, yet you being valuable and being valued by someone in particular aren't always the same thing.

But a more fundamental issue in these kinds of cases is that even when we can add our values together easily, incentives are aligned against agents restricting themselves when the chances are really low and they aren't confident others will do the same. It's easy to imagine: there's a very, very small chance you design a modified virus that *could* be mammalian transmissible, *could* translate to human transmission, *could* be released in a lab accident, and *could* seed an epidemic. There's a certainty, though, that if you don't do this research, you can't move on to other projects in your lab, get your next grant, or gain tenure, or whatever the risk might be. Your peers either aren't going to do GOF research, or they are, but they are unlikely to exercise restraint organically. So the incentive for *you* is to continue your work—after all, the chance that it will be your roll of the dice that leads to catastrophe is extremely small.

This issue becomes thornier in the contexts of the global governance of science. While the prominent examples are in the United States, even as recently as 2015, the funders in the United Kingdom, Netherlands, Japan, and China had all committed resources to GOF research. A typical worry, then, is that if one nation restricts or disincentivizes GOF research, then scientists will simply go elsewhere. But an additional worry is that meaningful regulation of GOF research, even the tiny number of studies that we expect will prompt the denial of funding or selection of an alternate experiment, is likely to be unpopular even for countries with a process. There is no shortage of literature out there claiming that advances in the life sciences are necessary for future US national security, and that do so in economic terms.[9] We might imagine that institutional interest in regulating virology, even if otherwise justified, will be impaired by the perception that "other countries don't play fair."

Global agreement on preventing catastrophic risks that arise from life sciences research is perhaps the thorniest issue here, but one that has been claimed as necessary since the Fink Report 20 years ago. There's a real need for global agreement not just on the fact that something needs to be done but on what exactly that looks like too.

And it likely needs to be larger than GOF research, which, as a species of dual-use research, is part of a larger whole of a small yet deeply concerning set of research.

The way forward here is likely to be a political process—broader and more comprehensive than the deliberative process of 2014. I say this because when our intuitions strongly diverge in biomedical ethics—stem cells, access to health care, climate change, and so on—we enter into political processes to resolve conflict. Those political processes can be ugly.

8

THE FUTURE OF GAIN OF FUNCTION

GOF is a live issue in science policy. As we've seen, despite a decade of debate over the proposed designs of regulation to manage the risks and benefits of this research appropriately, not much has changed. This has left proponents and critics of the research, in varying degrees, frustrated and dispirited. So the question then becomes, Where to next?

In this chapter, I discuss some of the next issues we are seeing develop around GOF. I'm going to start with what is perhaps the current elephant in the room in relation to the subject. That is the so-called lab-leak theory around the origins of COVID-19. I'm going to explain how the theory ties into the debate around GOF research, and while I can't rehearse every single move in what is now an incredibly broad and wide-ranging debate, I hope to show that the connection between the lab-leak theory and GOF is tenuous at best but it is doing so much work precisely because of the ambiguities we've been dealing with over the last few chapters.

The second issue I'll deal with arises in synthetic biology—the design of living organisms using advanced biotechnology—as a challenge to our current thinking and an opportunity for the future.[1] In this new regime, starting with a natural template or even a recombination of segments of a pathogen won't be necessary to create risky new organisms. That will require that we rethink how we govern science and security, and a warning about repeating the intractability of the last decade of debate about scientific governance.

The final thing I'll deal with is what GOF stands to teach us about larger questions of scientific pursuit. Even if we agree—as I do—that there are some GOF experiments that we ought not to pursue, maybe now or ever, the solution doesn't have to be an incredibly narrow variation on the status quo. GOF research, rather, is a chapter in a longer and broader story about how we do science, what we choose to study, and for whom science happens.

The Lab-Escape Theory

Any discussion about the origins of a pandemic virus, 4 years into a pandemic, is a work of probabilities. This might sound strange to optimists about science or those

GOF research, rather, is a chapter in a longer and broader story about how we do science, what we choose to study, and for whom science happens.

who have watched *Contagion*, but it is decidedly the case. The way that SARS-1 emerged (it's "proximal origins" as is used today in debates around COVID-19) took almost 10 years to decipher. The circumstances of the arrival of 2009 influenza A H1N1, colloquially termed "swine flu," are still murky. We believe it emerged from a reassortment of North American and Eurasian flu viruses in livestock, but where and when are still unknown. And the 1918 influenza pandemic, while commonly called "Spanish flu," was first detected in Kansas, though whether this was its actual origin point is and likely always will be unknown.[2] And while the suspected origins of the West African Ebola virus disease outbreak were confirmed in less than a year, that disease has almost no asymptotic spread, clear symptoms, and much slower movement than COVID-19 on account of no respiratory droplet nuclei transmission (i.e., it isn't "airborne") and only spreading after symptoms appear.[3]

Any claim made about the origins of COVID-19 that doesn't hedge, then, is irresponsible at best. Thus—and to put my cards on the table—while I maintain that the most likely explanation is a naturally occurring virus, there is little reason to settle completely on any single explanation. This will seem frustrating to those who have become convinced by one so-called side in this debate but, as we'll see, there's less to suggest the evidence for a laboratory origin of COVID-19 than appears on its face, and much of the argumentation for a spontaneous natural source is inference from previous outbreaks. In these kinds of debates, it is easy to slide from mere possibility to plausibility, but I am going to try to show how tenuous such a move would be.

To begin, we need to distinguish two separate elements to theories that hold COVID-19 originated from a laboratory in some capacity. The first part of these theories is that the SARS-CoV-2 virus has its origins in, or at the very least passed through, the activities of a laboratory. That laboratory is most often the Wuhan Institute of Virology (WIV), but it is sometimes noted that in a large biomedical hub like Wuhan, other labs exist, including the Wuhan Center for Disease Control. All theories that hold to a lab release include the premise, analytically so, that a lab release of COVID-19 involved, or began with, a laboratory.

There is a subset of theories, however, that maintain that in addition to emergence through a lab, SARS-CoV-2 was likely or even *necessarily* a virus that was subject to genetic modification. This is where GOF research enters the discussion: that SARS-CoV-2 may have derived from another (ostensibly less dangerous, these stories go) virus that was engineered to have greater transmissibility, virulence, or host range. There is a further, more extreme theory that such a modification was performed for biological weapons purposes but, with the exception of figures like Alex Jones or Senator Tom Cotton, most proponents of a lab-release theory view this further move as conspiratorial—so I'll set it aside.

What makes the idea of a lab release of some kind plausible to its proponents? A number of key pieces of evidence are commonplace. The first is that the market where 60 percent of the first reported cases of COVID-19 arose is located in the same city as and just a few miles from the WIV. This kind of proximity, proponents will say, is simply too surprising to be a coincidence.

The second piece that proponents will point to in this puzzle is the uncooperativeness of the Chinese government in providing access to samples, sequences, or labs. This often appeals to a number of different episodes in the development of the pandemic. First, there was the recalcitrance of the Chinese authorities to reveal information about the emergence of COVID-19. Then there was their unwillingness to allow WHO inspection teams into Wuhan and the WIV or to support them in their fact-finding missions. Then there was the work of Jesse Bloom in June 2021 suggesting that genetic COVID sequence data from the National Center for Biotechnology Information's Sequence Read Archive had allegedly been purged by the Chinese scientists.

A lack of confirmed animal reservoir and ecological history of the virus is also given as negative proof of an alternate theory to a naturally occurring pandemic—of which the lab leak is presumed to be the dominant contender. That is, while there are several possible animal hosts on the table right now, including the March 2023 hypothesis of a Chinese raccoon dog, there have been no animals found with older generations of COVID-19 in their system in the area or market.[4] Viruses evolve quickly, but the only sample we have of coronaviruses related to COVID-19 predates the virus by almost 40 years. Together, proponents claim, this provides a reason to believe that this virus lacks a point of transmission from animal to human, leading to the conclusion that some other start to the pandemic must have occurred.

The safety record of the Chinese lab is thrown into question too. Reporting early in the pandemic claimed that the Chinese government had, only months before the outbreak, issued a report and new biosafety guidelines that emerged from safety concerns in Chinese labs, and cited Washington diplomatic cables about those same safety cables.[5] This points to a standing issue that, to some, raises the possibility of a lab leak to a plausibility.

What of GOF research? The dominant piece of evidence on this front is a grant application to DARPA that was submitted in 2018 by a consortium including the EcoHealth Alliance, Ralph Baric at the University of North Carolina at Chapel Hill, and scientists at the WIV. That grant, which was ultimately denied, proposed to collect large numbers of samples of coronaviruses and then sequence the spike proteins of a large number of those viruses to determine which had the capacity to emerge into human-to-human transmission. Some supporters of the laboratory-leak theory point to this as evidence of motivation on the part of relevant scientists to engage in what they consider to be GOF research coronavirus experiments, and thus those scientists could have found funding elsewhere to continue their work. They cite further DARPA's rejection letter, which mentions that because the research involves potential GOF/DURC, if ultimately selected for funding, the contract for funding would have to include a risk-mitigation plan.

Most of the claims above are plausible in the sense that the basic events seem to add up. Yet the events themselves aren't sufficient to demonstrate anything about anything; there are pretty specific trains of thought one has to roll down to take these from facts to evidence of a lab leak, much less connecting them all together to make a compelling theory. Individually, there are a number of easy responses to question the strength of the conclusions of proponents of a lab-leak theory. The first is that proximity is simply not a measure of culpability by itself in something like a lab release. For reference, the Hendra virus outbreak in Queensland, Australia, in the 1990s started in a horse paddock roughly the same distance from the University of Queensland's virology laboratories as the Wuhan seafood market is from the WIV. But never has it been claimed in any credible source (much less, say, in the *New York Times* or *Vanity Fair*) that the outbreak in Australia that killed 19 horses and a human could have emerged, even possibly, from a lab.

Second, the uncooperativeness of the Chinese government is explainable in several other ways, all of which are at *least* as plausible as a theoretical effort to cover up a lab leak. For example, as an anonymous Hong Kong virologist noted during the height of the online debate about COVID-19, the Wuhan seafood market is owned by the daughter of Yu Zhusheng, a prominent Chinese business mogul with close ties to the Hubei Province Communist Party chapter.[6] It is entirely possible—as is common in countries worldwide—that the wildlife trade in the markets of Wuhan was considerably more developed than has previously been thought and may even be one of a number of illicit activities conducted under the auspices of powerful Chinese business figures. In this case, the face-saving game is plausibly less about the laboratories in Wuhan than it is about embarrassment for a regional branch of the Communist Party; a thorough investigation of Wuhan would not reveal a lab leak but rather impact the party in Hubei Province in a much more serious way connected to its trade.

Another and I think equally plausible theory (by the standards of the lab-leak theory) is that the Chinese government suffered a blow to *miànzi* or "face," its standing, in the way senior US officials reacted to the early stages of the pandemic. Between Cotton above referring to the pandemic as plausibly caused by a biological weapon, President Donald Trump's repeated use of the term "China virus," and Mike Pompeo's corresponding "Wuhan virus" (which some of my students use to this day), it is quite plausible that the Chinese simply *do not want to cooperate*.[7] And being a nuclear-tipped superpower, they do not have to. As someone who grew up in a middle power, I am used to embarrassed superpowers taking their toys and going home—including the superpower I currently live and hold citizenship in—but strangely, citizens born into frequently embarrassed superpowers seem to miss this behavior in their own nation and hyperfocus on it elsewhere.

Of course, this is speculation—but speculation based on positive evidence rather than on the absence of it, such as the unavailability of laboratory records. The central point here is that there is no reason to believe *one way or the other* that the uncooperativeness of China can be tied to a lab leak. Nor is this exonerative of the Chinese government; any uncooperativeness, whatever the reason, harms a search for the origins of COVID-19, natural or otherwise. For instance, recent allegations that the sequence for COVID-19 appeared in the queue for submission to the gene repository GenBank two weeks before the WHO received it portray a Chinese government potentially reluctant to share critical genetic information about the virus with the world.[8] But that isn't effective evidence of a lab leak (and that it was submitted at all is arguably evidence to the contrary). No one is impressed with the Chinese Communist Party, but that gives no reason to favor one theory over any other.

The lack of an animal reservoir is tied in part to the cleansing of the seafood market, which may be tied to the above uncooperativeness. But it's also worth remembering what I began with: the discovery of the proximate origins of SARS-1 took almost a decade. Given that scientific debate about the laboratory origins of SARS-CoV-2 took less than a year to start, it might simply be that the public and scientists alike are just woefully optimistic about what it actually takes to identify the origins of a pandemic disease.

The biosafety charges against the Chinese are interesting but, again, not positive evidence of a lab leak. It seems ill-timed, yet a closer inspection of Chinese governance of the life sciences reveals a different picture. The few comprehensive English-language descriptions of Chinese biosciences laws identify efforts to develop Chinese biological safety and security guidelines since the 1980s. In 2018, these came to a head with the thirteenth People's Conference designating the biosafety/biosecurity legislation as one of the class-three legislative projects, requiring further deliberation and launching a process that would conclude with new laws coming into force in 2021—the same year that the alleged diplomatic cables arrived from

Washington with concerns about WIV's safety record.[9] At the same time, multilateral meetings in Tianjin around a code of ethics for biologists began in 2018, resulting in the Tianjin Biosecurity Guidelines for Codes of Conduct being released, again, in 2021.[10]

There have been lab accidents in China in the last 20 years, including famously 2 accidents involving SARS-1 in 2004. But there have also been significant numbers of lab accidents in the United States and United Kingdom, and there is no evidence to suggest that the rate is proportionally higher in China than it is elsewhere.[11] Nor does movement on biosafety regulations signify an egregious gap in Chinese governance relative to the rest of the world. Like many areas of the GOF debate, this kind of claim can be attributed to people "discovering" issues during COVID that others have spent decades on and misreading routine or even praiseworthy actions as signs of suspicion.

None of which is to say that China does not deserve condemnation for its treatment of the true, original whistleblowers for COVID-19: scientists who were imprisoned and even killed to suppress information of the virus.[12] Nor is it the case that Chinese authorities should be excused for stymieing the work of investigators in Wuhan.[13] But I note that in terms of causes, lab-leak proponents have not yet addressed the fact that the reason we have no international verification mechanism for high-containment laboratories and other potential sites of biological weapons activities is the United States, which axed a proposal to do just that on the eve of the Fifth Review Conference of the Biological Weapons Convention. Nor have lab-leak proponents placed blame for China's withdrawal at the feet of Trump and those like him; apparently, when people in the United States must suffer deranged comments from a president with a shrug, other sovereign nations must do the same.

What of the GOF theory? The DARPA grant has been described by futurist Jamie Metzl in the following way: "If I applied for funding to paint Central Park purple and was denied, but then a year later we woke up to find Central Park painted purple, I'd be a prime suspect. If I hid the history of my grant application while leading a campaign to label anyone asking common-sense questions about how this may have happened as a conspiracy theorist, I'd be a fraud."[14] But there's no evidence this grant was hidden. There was no call of which I know for all researchers who may or may not have applied for a grant related to SARS-CoV viruses to unearth their old applications. And if it's too pie-in-the-sky for the blue-sky arm of the US Department of Defense, there's every likelihood it simply didn't go ahead. Importantly, and as proponents of GOF are quick to remind us, this kind of research is expensive to do. Of course, we don't know that for sure, but nothing follows—*nothing* follows—from the existence of a grant application in 2018 to anything else.

There's another wrinkle in the story, however, for anyone who took the time to read that grant. The grant *does* mention GOF, but in the negative sense. This is because the proposed research involving the spike proteins is explicitly stated to be conducted at the University of North Carolina using non-GOF methods. The authors then cite a paper by Ralph Baric's lab at that university that involves the addition of spike proteins to a modified SARS-like virus designed for replication in mice.[15] The paper in question is quite reflective on its relationship to GOF research, and while the mouse viruses are not human-type viruses and are attenuated relative to wild SARS, one can infer increases in pathogenicity from these experiments. Yet the research itself is not the kind that involves the creation of human-transmissible viruses. This paper has been long misidentified as GOF research but remains something quite different—it uses attenuated mouse strains, the kind of research that has been advocated as an alternative to GOF for flu research.

Finally, and perhaps most significant, the work was proposed to DARPA in March 2018. It's important here to understand the scientific process. Getting rejected by DARPA, finding a new funder with the appropriate grant stream, reapplying, being reviewed, getting approved, and actually receiving the money to *go into the wilds of China and find bats to sample*, and then getting the work done at the WIV (even though the piece of research that was allegedly concerning was going to be done at the home of the world experts on this area of the science, Baric's lab at the University of North Carolina) in 18 months to the extent that it could lead to a lab release seeding a global pandemic is implausible on its face. Anyone reading this with even the slightest inkling of how the life sciences works will note that actually going out and getting the samples from bats is a huge, rate-limiting step. And given the task involved in collecting and processing thousands of bat samples, the likelihood that the researchers stumbled across one that would produce COVID and then—quite out of step with the proposal itself—chose to create a human virus instead of a mouse virus sounds like something out of *28 Days Later*.

This, then, is the problem with these theories. Each and every proposition could be true: China is recalcitrant, there's no confirmed animal reservoir, and so on. But none of them seem to hold in context, and certainly they tell nothing that provides us an inference from which we can establish that a lab leak is plausible or likely. The only evidence, moreover, that this is GOF research is a single grant application that was not only submitted to one of the only funders in the world to truly fund blue-sky research but also purports to have been able to be turned around and conducted prior to December 2019 and through means far beyond what was actually proposed. It doesn't mean a lab leak is *impossible*, but possibility and plausibility are quite different things.

Yet in some ways, the damage is already done. Two preprints have sparked controversy during COVID in predictable ways. In 2022, a Boston University preprint made headlines when it described a recombination study utilizing an older version of SARS-CoV-2 and the at-the-time circulating Omicron variant to find out which piece of the Omicron spike protein caused the new current lineage to override vaccine protections in humans. While the study abstract did report an 80 percent death rate in ACE2 mice—the mice used in COVID-19 research that led to, among others, the mRNA vaccines so many people in the United States have received—and sounded like the COVID-19 version of the mousepox study, we should note a couple of things. First, the mice were exposed to extremely high rates of virus in their noses—more than a standard infectious dose. And that 80 percent was actually less than the old, wild virus, which killed 100 percent of those mice. We know that no COVID-19 virus has ever killed 100 percent of its hosts, so we have absolutely no reason to believe that a weaker strain in purpose-bred mice is going to be worse than the real thing.[16] This only looks like GOF research on its face.

Another breathless headline made similar claims about a "100% lethal" version of COVID-19 made in China.[17] In this case, the virus was a coronavirus but wasn't related to COVID-19; it was a virus found in pangolins. The virus, cataloged initially in 2018, had mutated in the lab inside the cells it was used in.[18] Researchers attempted to characterize the virus they ended up with rather than the virus they thought they had. And while it did kill all four mice they inserted it into, they were using an unproven, new mouse "brand," and in previous studies the lethality had never been observed.[19] Moreover, the virus in question isn't known to infect humans, though in experiments it has infected other mammals.[20]

So, again, these aren't GOF studies in the sense we're talking about. The last one isn't even GOF at all but just an exploration of a virus the researchers had already studied. The problem is that when you're inoculated to believe that *everything* is GOF and are worried about GOF, you're worried about anything! This is the move spoken about in chapter 1 taken on new life. And that life is extremely detrimental to any discussion about science policy.

We should strive to avoid, wherever possible, the contention that COVID came from a lab as a stand-in for a data point on the risks of GOF. So far, there is no credible evidence that GOF is involved in any way in even the most plausible version of a lab-leak theory, much less any more robust theory of the origins of COVID-19. In a field that has few empirical data points on which to build policy, it is really tempting to hold up COVID-19 as an example of what could happen from GOF research. But this both oversells the relationship of COVID-19 to GOF and undersells the risks of GOF research itself. After all, GOF research, the kind about which we are concerned, stands to be considerably more dangerous—by definition—than the emergence of a wild virus.

COVID-19's death toll reflects an outrageous policy failure, and I have argued that the response to that pandemic is an extreme injustice visited on us by our leaders.[21] Still, it isn't what we would or should expect from GOF. That's what makes GOF so scary and the stakes so high for the governance of science. But inserting a naturally occurring pandemic, and the policies we need to repair our public health systems from that ongoing calamity and prepare for the next, into another debate is neither productive nor useful. We should resist this narrative, unless faced with evidence of a more compelling kind, and rely on fewer stretches of imagination when the less cinematic but more plausible—and in some ways, more confronting—reality is that we as a globe simply failed in the face of biological inevitability.

Synthetic Biology

GOF research is, in some sense, a synthetic biological endeavor as it involves the creation of novel biological "constructs." Yet it is, as we've discussed, not typically subject to the rational design principles that we would associate with engineering. Aqueducts obey incredibly regimented principles that have been codified for thousands of years; designing viruses is considerably more error prone.

Over the coming decades, however, synthetic biology promises the following. There is a future in which the genetic determinants of viruses are well understood and the interactions of genes that currently frustrate virologists and other life scientists are known. At that point, the possibilities of what we might call "genetic space" will be easier to explore, including biological organisms that not only have never occurred in nature like the existing GOF creations but also have no natural analogue at all.

This is a huge challenge for dual-use research in general, but importantly, the current regulations in the United States and indeed most of the rest of the world do not have a handle on this problem. The challenge is that most regulations begin with a list of pathogens that are "of concern." The change to research categories 1 and 2 attempts to overcome this challenge, but the degree to which it succeeds has yet to be seen.

For a long time, policymakers have recognized that lists of viruses by and large are poor ways of setting policy in virology. For example, restricting the use of coronaviruses would be a poor move on account of almost all of those viruses being harmless or nonlethal, yet the list could not have been updated fast enough to allow someone to do dual-use research on COVID early in the pandemic.

But synthetic biology poses another challenge. Because it involves the creation of novel constructs, there is potentially no way to regulate synthetic biology like we do traditional virological research. If synthetic biology lives up to its promise, then eventually researchers will not require an original, live template to work it; rather, they will engage in the rational design and synthesis of viruses from their constituent DNA or RNA. This presents a challenge because new GOF research might not involve wild viruses that are modified but instead the sequencing of novel viruses from constituent parts.

This is made all the more worrisome through the rapidly reducing costs of research in synthetic biology, which are understood to obey something similar to Moore's law: every 18 months, the cost of synthesis of genetic constructs halves. This means that over the coming decades, the cost of synthesizing new biological constructs could become trivial. What this means in practice is that someone could begin a process of optimizing the flu virus for lethality and transmissibility and push it to within the limits of biological possibility. Or they could simply design a totally new, never-before-seen pathogen.

This is still some distance away, I suspect. People have been worrying about synthetic biology for almost 20 years, but the fears have yet to be realized. This is due less to the enthusiasm of the community in making scientific progress and more to the financial incentives that encourage pharmaceutical development over basic research along with the sheer complexity of biology.

In policy terms, though—and remembering that GOF is a policy idea—another 20 or 30 years is not a long time at all. Recall that even the cosmetic changes in moving from the initial H5N1 policy in 2013 through to P3CO took half a decade; the subtle updates to P3CO in the 2024 dual-use policy took a similar time frame. A more substantive change is exponentially more difficult. Complaints about the current regulations of harmful pathogens and risky research are more than 20 years old at this stage, and yet no better solution has been found.

One solution, as I've discussed in the context of ferrets, is to change how we fund scientific research. That is, much of the issue that arises in synthetic biology is a lack of infrastructure to realize its benefits. This makes the risk-benefit ratio skewed in favor of risks, but with a strong degree of uncertainty.

But this isn't sufficient for synthetic biology, in which the power to develop novel pathogens will be extremely distributed. What is needed, I suspect, is international regulation and diplomacy, but that is currently lacking. Fortunately, we have the time to rebuild those connections and develop the power to secure the life sciences against misuse—and develop the infrastructure to respond to the outbreaks that may arise.

The central issue, then, is that outbreaks from synthetic biology are plausible. Nevertheless, I think we should be careful about our estimates of how frequently these will occur. The answer to how often will determine, however, whether we should maintain a model of security or move to something more like harm reduction—empowering people to take risks responsibly and providing them the means to do so.

This hasn't been explored in the literature and thus is too far afield for this book. But it is worth noting, as in previous chapters, that the traditional security model of response to risks in biology may not be apt. After all, if we think biology is useful—and it seems to be—the policies we choose might need to be better informed as a way to mitigate risk and respond to harm, which are closer to classical public health. That is the crux of the debate around GOF research now and in the future.

For Whom Are We Researching?

GOF research is ostensibly pursued for the public good. As we've seen, however, there are serious questions about the magnitude of that benefit, how we weigh the benefit against its risks, and what we do in cases where the risks and benefits aren't distributed equally. This isn't a new problem for science, and in some ways it is a characteristic of a modern scientific landscape that is neither well understood nor open to popular understanding in modern politics. But with GOF research, these old challenges are demonstrably more high stakes as the public becomes more suspicious of the activities of those life scientists who work with pathogens.

How do we bridge this gap? One standard account is that we ought to educate more people. This is almost certainly true on its own—education is a public good in its own right—but it doesn't easily capture the problem of GOF. That's because opposition to GOF research, or support of it, isn't strictly dependent on education or knowledge. The contours of the debate in professional science mean that, even if scientists are sometimes wrong about specific scientific debate, nothing rides on a knowledge of virology that lends itself to taking one side or the other. Certainly, coronavirus or flu virologists *tend* to be more supportive of GOF research. They aren't always, though, and in any case, extremely educated opponents exist who have more than a sufficient understanding of the science to form an opposing view.

The deeper problem is that the *values* people bring to science shape our understanding of the significance of GOF. It's true that many virologists view GOF research, or at least GOF research that is also virology research, as an essential part of an effective public health response. Still, as scholars have noted for the better part of 40 years, developed nations are increasingly trained to view basic biology research as not just important but also uniquely critical to pandemic preparedness. As public health infrastructure has decayed, communities have become increasingly alienated from the promotion of their collective health, and medical care is increasingly stratified between the haves and have-nots, we should reflect on whether the reason the public is so insensitive to claims of the benefits of GOF is because people are, with good reason, increasingly skeptical of their relationship to science along with the prime role that state—not democratic—interests play in shaping it.

This is far from a new claim about science, but little has been said about it in the context of GOF research. Yet as the movement—right or wrong—to connect GOF research to the ongoing COVID-19 pandemic gains steam, scientists and their defenders are left with a deep problem. Claims like Rand Paul's, that individual scientists are simply engaged in irresponsible or even criminal acts, are easy to frame in his own libertarian worldview, which doesn't admit to collective action, much less action on the part of *institutions* or organs of the state. It's an easy perspective; to riff off former UK prime minister Margaret Thatcher, when there are no institutions and only individuals, you don't need to have a particularly complex theory of collective action.

But there isn't a well-defined, coherent message operating that can offer serious competition to this theory. In point of fact, in the dual-use debate the overarching account asserts the primacy of the *individual freedom of the scientist*. (To wrap this book in a neat bow, Jo Husbands first pointed this out to me in a rare, quiet moment chatting on the margins of a meeting of the Biological Weapons Convention.) That's more or less the account folks like Senator Paul rely on! How science connects to public values, as they exist today, for real people is justified under something like a "trickle-down epistemology": knowledge created by scientists eventually finds its way to the common person, usually in the form of medicine, technology, and so on.

Yet that's not how it has to be. Rather than mount an independent argument on this, I want to refer back to a much older, foundational document in US science policy: *Science: The Endless Frontier*, authored by engineer Vannevar Bush in 1945 in consultation with a number of committees composed of early twentieth-century scientific luminaries. References to this document in science policy focus on its claim either that science is essential to "our security as a nation, to our better health, to more jobs," or that basic research without specific practical aims is "the pacemaker of technological progress."[22]

What is often missed, though, is one of the few verbatim quotes in that report from a committee chaired by

Henry Allen Moe, who was not a scientist but instead a humanist and president of the Guggenheim Foundation. In answering a question from President Franklin Delano Roosevelt, "Can an effective program be proposed for discovering and developing scientific talent in American youth so that the continuing future of scientific research in this country may be assured on a level comparable to what has been done during the war?" the Moe committee replied,

> As citizens, as good citizens, we therefore think that we must have in mind while examining the question before us—the discovery and development of scientific talent—the needs of the whole national welfare. We could not suggest to you a program which would syphon into science and technology a disproportionately large share of the nation's highest abilities, without doing harm to the nation, nor, indeed, without crippling science.... Science cannot live by and unto itself alone.[23]

That is, the pursuit of science for its own sake is important, but it is tied to the health of the nation as a whole too. Right now, proponents of GOF research cannot, as we've seen multiple times, offer a coherent account of why a particularly risky kind of research *is a risk worth taking on behalf of everyone*. There isn't a publicly justifiable account that nonscientists have bought into about the relationship between GOF research and the individual or public good. It's rational for people to refuse to buy into a system that requires them to accept risk with no clear path to their own interests.

As I said, this is an issue that all science faces to a greater or lesser extent. What makes GOF research distinct is that it imposes what appears to be a serious but hard to capture *risk* on people. It isn't a mere disagreement about the value of science for its own sake or—infamously—whether tax dollars should go to studies involving shrimp on treadmills. This is a risk of harm that individuals are being asked to accept without any payoff. And that is, to many, unacceptable.

GLOSSARY

Dual-use research
Scientific research or its applications that have the potential to be used to help or harm humanity.

Dual-use research of concern (DURC)
A policy term used by the NSABB and others to describe "research that, based on current understanding, can be reasonably anticipated to provide knowledge, products, or technologies that could be directly misapplied by others to pose a threat to public health and safety, agricultural crops and other plants, animals, the environment, or material."

Federal advisory committee
Committees that provide the US president and executive branch with advice on a range of topics. They are governed by the Federal Advisory Committee Act of 1972. About 1,000 federal advisory committees exist in the US government.

Fink Report
The first report on dual-use research published in 2003, formally titled *Biosecurity Research in an Age of Terrorism*. Named after its chair, Gerald Fink, it was succeeded by *Globalization, Biosecurity, and the Future of the Life Sciences*, or the "Lemon-Relman Report," in 2007.

Gain-of-function (GOF) research
A scientific study that is reasonably expected to create a pathogen (bacteria, fungus, etc.) with an enhanced virulence, transmissibility, or host range from its natural template to such an extent that it poses a large-scale threat to human, animal, or plant populations.

Information hazard
A risk that arises from the dissemination of (true) information that may cause harm or enable some agent to cause harm.

National Science Advisory Board for Biosecurity (NSABB)
Originally outlined in the Fink Report as the National Science Advisory Board for Biodefense, the NSABB is a federal advisory committee that provides recommendations to the US National Institutes of Health and US government on issues pertaining to dual-use, GOF, biosecurity, and biosafety regulations.

Potential pathogen care and oversight (P3CO)
Part of the title of a policy guidance document issued by the US White House Office of Science and Technology Policy titled *Recommended Policy Guidance for Departmental Development of Review Mechanisms for Potential Pandemic Pathogen Care and Oversight*.

Serial passaging
The microbiological process of taking a pathogen and passing it through a host repeatedly. These hosts may be cell cultures or animals such as ferrets.

NOTES

Chapter 1

1. There are lots of great articles on the pros and cons of paradigm cases. One that stands out is John D. Arras, "Getting Down to Cases: The Revival of Casuistry in Bioethics," *Journal of Medicine and Philosophy* 16, no. 1 (1991): 29–51, https://doi.org/10.1093/jmp/16.1.29.

2. For a conception of American Association for the Advancement of Science as a trade union/organization of sorts that permits scientists to engage in collective bargaining with the US federal government in particular, see Daniel J. Kevles, *The Physicists: The History of a Scientific Community in Modern America* (Cambridge, MA: Harvard University Press, 1995).

3. Martin Enserink, "Scientists Brace for Media Storm around Controversial Flu Studies," *Science Insider*, November 23, 2011, https://www.science.org/content/article/scientists-brace-media-storm-around-controversial-flu-studies.

4. Ander Herfst et al., "Airborne Transmission of Influenza A/H5N1 Virus between Ferrets," *Science* 336, no. 6088 (June 2012): 1534–1541, https://doi.org/10.1126/science.1213362.

5. For an explanation of the "real" case fatality rate, see F. C. K. Li et al., "Finding the Real-Case-Fatality Rate of H5N1 Avian Influenza," *Journal of Epidemiology and Community Health* 62, no. 6 (June 2008): 555–559, doi:10.1136/jech.2007.064030.

6. Masaki Imai et al., "Experimental Adaptation of an Influenza H5 HA Confers Respiratory Droplet Transmission to a Reassortant H5 HA/H1N1 Virus in Ferrets," *Nature* 486, no. 7403 (2012): 420–428, https://doi.org/10.1038/nature10831.

7. Anna Cornelia Nieuwenweg et al., "Emerging Biotechnology and Information Hazards," in *Emerging Threats of Synthetic Biology and Biotechnology*, ed. Benjamin D. Trump et al. (Dordrecht: Springer, 2021), 131–140, https://doi.org/10.1007/978-94-024-2086-9_9.

8. National Research Council, *Potential Risks and Benefits of Gain-of-Function Research: Summary of a Workshop* (Washington, DC: National Academies Press, 2015).

9. For the interested, the NLM MeSH terms used are "viruses," "virus," and "virology"; the search string thus looks like: (((viruses[MeSH Terms]) OR (virus[MeSH Terms])) OR (virology[MeSH Terms])) AND ("gain of function").

For those really interested, when you search the same on the PubMed Commons Archive, you get 3,110 articles mentioning GOF and virology. So a lot more alleged GOF papers than before! The number of articles that mention *any* GOF, however, is not 18,000 but rather *100,000*. So in raw terms, the proportion of GOF virology articles to all GOF doesn't change: only about 1 in 5.5 articles that use the term concerns viruses. Nevertheless, most of that happened after the controversy began. Of the PubMed Commons sample between 1982 and 2011, 105,000 virology papers are archived, 771 of which mention GOF. This means that approximately 0.74 percent, or 1 in 136, of virology papers mention GOF in any sense of the word—broad or narrow.

10. Victor Bertuzzi and Stefano DiRita, "Gain-of-Function Research Advances Knowledge and Saves Lives," *STAT* (blog), December 23, 2021, https://www.statnews.com/2021/12/23/gain-of-function-research-advances-knowledge-and-saves-lives/.

11. Felicia Goodrum et al., "Virology under the Microscope—a Call for Rational Discourse," *mBio* 14, no. 1 (January 26, 2023), https://journals.asm.org/doi/10.1128/mbio.00188-23.

12. Caroline Schuerger et al., "Understanding the Global Gain-of-Function Research Landscape," Center for Security and Emerging Technology, August 2023, https://cset.georgetown.edu/publication/understanding-the-global-gain-of-function-research-landscape/.

13. Marc Lipsitch, "Why Do Exceptionally Dangerous Gain-of-Function Experiments in Influenza?," in *Influenza Virus*, ed. Yohei Yamauchi (New York: Humana Press, 2018), 589.

14. For a minor pilot point in this regard, see Ian Irvine, *The Last Albatross* (New York: Simon and Schuster, 2000).

Chapter 2

1. See, for example, Nicholas Greig Evans, "Ethical and Philosophical Considerations for Gain-of-Function Policy: The Importance of Alternate Experiments," *Frontiers in Bioengineering and Biotechnology* 6 (February 2018): e1875, https://doi.org/10.3389/fbioe.2018.00011; Nicholas Greig Evans, "Great Expectations—Ethics, Avian Flu and the Value of Progress," *Journal of Medical Ethics* 39, no. 4 (March 2013): 209–213, https://doi.org/10.1136/medethics-2012-100712; Nicholas Greig Evans, Marc Lipsitch, and Meira Levinson, "The Ethics of Biosafety Considerations in Gain-of-Function Research Resulting in the Creation of Potential Pandemic Pathogens," *Journal of Medical Ethics* 41, no. 11 (November 2015): 901–908, https://doi.org/10.1136/medethics-2014

-102619; Nancy Connell, "Immunological Modulation," in *Innovation, Dual Use, and Security*, ed. Jonathan B. Tucker (Cambridge, MA: MIT Press, 2012), 187–198; Luke Kemp et al., "Point of View: Bioengineering Horizon Scan 2020," *eLife* 9 (May 29, 2020): e54489, https://doi.org/10.7554/eLife.54489; Jo L. Husbands, "The Challenge of Framing for Efforts to Mitigate the Risks of 'Dual Use' Research in the Life Sciences," *Futures* (March 2018), https://doi.org/10.1016/j.futures.2018.03.007.

2. National Science Advisory Board for Biosecurity, *Proposed Framework for the Oversight of Dual Use Life Sciences Research: Strategies for Minimizing the Potential Misuse of Research Information* (Bethesda, MD: National Institutes of Health, June 2007), https://osp.od.nih.gov/wp-content/uploads/Proposed-Oversight-Framework-for-Dual-Use-Research.pdf.

3. For historical comparisons, see Nicholas Greig Evans, "Contrasting Dual-Use Issues in Biology and Nuclear Science," in *On the Dual Uses of Science and Ethics*, ed. Michael J. Selgelid and Brian Rappert (Canberra: ANU Press, 2013), 255–273, http://press.anu.edu.au/wp-content/uploads/2013/12/ch161.pdf.

4. Seumas Miller and Michael J. Selgelid, *Ethical and Philosophical Consideration of the Dual-Use Dilemma in the Biological Sciences* (Dordrecht: Springer, 2008).

5. Christian Enemark, "United States Biodefense, International Law, and the Problem of Intent," *Politics and the Life Sciences* 24, no. 1–2 (2005): 32–42.

6. Charles Coulston Gillispie, "Science and Secret Weapons Development in Revolutionary France, 1792–1804: A Documentary History," *Historical Studies in the Physical and Biological Sciences* 23, no. 1 (1992): 35–152, https://doi.org/10.2307/27757692.

7. Frank Fenner and Bernardino Fantini, *Biological Control of Vertebrate Pests: The History of Myxomatosis, an Experiment in Evolution* (Wallingford, UK: CABI Publishing, 1999).

8. Ronald J. Jackson et al., "Infertility in Mice Induced by a Recombinant Ectromelia Virus Expressing Mouse Zona Pellucida Glycoprotein 3," *Biology of Reproduction* 58, no. 1 (January 1998): 152–159, doi:10.1095/biolreprod58.1.152.

9. Joseph Tartal, *Corrective and Preventive Action Basics* (Silver Spring, MD: US Food and Drug Administration, 2014), https://www.fda.gov/files/about%20fda/published/CDRH-Learn-Presenation--Corrective-and-Preventive-Action-Basics.pdf.

10. Michael J. Selgelid and Lorna Weir, "The Mousepox Experience," *EMBO Reports* 11, no. 1 (January 2010): 18–24, https://doi.org/10.1038/embor.2009.270.

11. Jeronimo Cello, Aniko V. Paul, and Eckard Wimmer, "Chemical Synthesis of Poliovirus CDNA: Generation of Infectious Virus in the Absence of Natural Template," *Science* 297, no. 5583 (2002): 1016–1018.

12. Michael J. Selgelid and Lorna Weir, "Reflections on the Synthetic Production of Poliovirus," *Bulletin of the Atomic Scientists* 66, no. 3 (2010): 1–9, https://doi.org/10.2968/066003001.

13. Andrew Pollack, "Scientists Create a Live Polio Virus," *New York Times*, July 2, 2002.

14. National Research Council, *Biotechnology Research in an Age of Terrorism* (Washington, DC: National Academies Press, 2004), https://doi.org/10.17226/10827.

15. Gerald L. Epstein, "Controlling Biological Warfare Threats: Resolving Potential Tensions among the Research Community, Industry, and the National Security Community," *Critical Reviews in Microbiology* 27, no. 4 (2001): 321–354.

16. National Research Council, *Biotechnology Research in an Age of Terrorism*, 5–6; emphasis in original.

17. Scott Shane and Eric Lichtblau, "Scientist Is Paid Millions by U.S. in Anthrax Suit," *New York Times*, June 28, 2008, Washington, https://www.nytimes.com/2008/06/28/washington/28hatfill.html.

18. Laura Spinney, *Pale Rider: The Spanish Flu of 1918 and How It Changed the World* (New York: PublicAffairs, 2017).

19. Jeffery K. Taubenberger et al., "Characterization of the 1918 Influenza Virus Polymerase Genes," *Nature* 437 (7060): 889–893, doi:10.1038/nature04230; Terrence M. Tumpey et al., "Characterization of the Reconstructed 1918 Spanish Influenza Pandemic Virus," *Science* 310, no. 5745 (2005): 77–80, doi:10.1126/science.1119392.

20. "The Deadliest Flu: The Complete Story of the Discovery and Reconstruction of the 1918 Pandemic Virus | Pandemic Influenza (Flu)," CDC, 2019, https://archive.cdc.gov/www_cdc_gov/flu/pandemic-resources/reconstruction-1918-virus.html.

21. Nicholas G. Evans, Michael J. Selgelid, and Robert Mark Simpson, "Reconciling Regulation with Scientific Autonomy in Dual-Use Research," *Journal of Medicine and Philosophy: A Forum for Bioethics and Philosophy of Medicine* 47, no. 1 (February 1, 2022): 72–94, https://doi.org/10.1093/jmp/jhab041.

Chapter 3

1. Sander Herfst et al., "Airborne Transmission of Influenza A/H5N1 Virus between Ferrets," *Science* 336, no. 6088 (2012): 1534–1541, doi:10.1126/science.1213362.

2. Herfst et al., "Airborne Transmission of Influenza A/H5N1 Virus between Ferrets."

3. Masaki Imai et al., "Experimental Adaptation of an Influenza H5 HA Confers Respiratory Droplet Transmission to a Reassortant H5 HA/H1N1 Virus in Ferrets," *Nature* 486, no. 7403 (2012): 420–428, doi:10.1038/nature10831.

4. Kathryn A. Radigan et al., "Modeling Human Influenza Infection in the Laboratory," *Infection and Drug Resistance* 8 (2015): 315, https://www.ncbi.nlm.nih.gov/pmc/articles/PMC4560508/.

5. Jessica A. Belser, Jacqueline M. Katz, and Terrence M. Tumpey, "The Ferret as a Model Organism to Study Influenza A Virus Infection," *Disease Models and Mechanisms* 4, no. 5 (2011): 575, https://www.ncbi.nlm.nih.gov/pmc/articles/PMC3180220/.

6. Michael R. Dietrich et al., "How to Choose Your Research Organism," *Studies in History and Philosophy of Biological and Biomedical Sciences* 80 (2020): 101227, https://pubmed.ncbi.nlm.nih.gov/31883711/.

7. Jessica A. Belser et al., "A Guide for the Use of the Ferret Model for Influenza Virus Infection," *American Journal of Pathology* 190, no. 1 (2020): 11–24, doi:10.1016/j.ajpath.2019.09.017.

8. Ding Y. Oh and Aeron C. Hurt, "Using the Ferret as an Animal Model for Investigating Influenza Antiviral Effectiveness," *Frontiers in Microbiology* 7 (February 2016): 80, doi:10.3389/fmicb.2016.00080.

9. Herfst et al., "Airborne Transmission of Influenza A/H5N1 Virus between Ferrets."

10. Imai et al., "Experimental Adaptation of an Influenza H5 HA Confers Respiratory Droplet Transmission to a Reassortant H5 HA/H1N1 Virus in Ferrets."

11. Imai et al., "Experimental Adaptation of an Influenza H5 HA Confers Respiratory Droplet Transmission to a Reassortant H5 HA/H1N1 Virus in Ferrets."

Chapter 4

1. Martin Enserink, "Scientists Brace for Media Storm around Controversial Flu Studies," *Science Insider*, November 23, 2011, https://www.science.org/content/article/scientists-brace-media-storm-around-controversial-flu-studies.

2. Katherine Harmon, "What Really Happened in Malta This September When Contagious Bird Flu Was First Announced?," *Scientific American*, December 30, 2011.

3. Enserink, "Scientists Brace for Media Storm around Controversial Flu Studies."

4. Denise Grady and William J. Broad, "Seeing Terror Risk, U.S. Asks Journals to Cut Flu Study Facts," *New York Times*, December 20, 2011, Health, https://www.nytimes.com/2011/12/21/health/fearing-terrorism-us-asks-journals-to-censor-articles-on-virus.html.

5. Joe Palca, "Panel Seeks to Safeguard Biological Research," NPR, July 2, 2005, Science, https://www.npr.org/templates/story/story.php?storyId=4727492; Bruce Alberts, "Modeling Attacks on the Food Supply," *Proceedings of the National Academy of Sciences of the United States of America* 102, no. 28 (July 12, 2005): 9737–9738, https://doi.org/10.1073/pnas.0504944102.

6. Howard Morland, "Born Secret," *Cardozo Law Review* 26, no. 4 (2005): 1401–1408.

7. "Press Statement on the NSABB Review of H5N1 Research," National Institutes of Health, September 18, 2015, https://www.nih.gov/news-events/news-releases/press-statement-nsabb-review-h5n1-research.

8. The WHO website changes constantly, but this statement is archived in a National Research Council Report on the H5N1 controversy. See National Research Council, "Official Statements," in *Perspectives on Research with H5N1 Avian Influenza: Scientific Inquiry, Communication, Controversy: Summary of a Workshop* (Washington, DC: National Academies Press, 2013), 55–80, https://www.ncbi.nlm.nih.gov/books/NBK206979/.

9. Kenneth I. Berns et al., "Adaptations of Avian Flu Virus Are a Cause for Concern," *Science* 335, no. 6069 (2012): 660–661, doi:10.1126/science.1217994.

10. Donald Kennedy, "Editorial: Better Never than Late," *Science* 310, no. 5746 (2005): 195.

11. Jeremy Youde, *Global Health Governance* (Cambridge, UK: Polity, 2012).

12. Paul Berg et al., "Summary Statement of the Asilomar Conference on Recombinant DNA Molecules," *Proceedings of the National Academy of Sciences of the United States of America* 72, no. 6 (June 1975): 1981–1984, https://doi.org/10.2307/64627?ref=search-gateway:7916755f3aea385da8021f36d2ebd951.

13. Peter Palese, "Don't Censor Life-Saving Science: Controlling Who Is Allowed Access to Information about Mutations in the H5N1 Bird Flu Virus Is Unacceptable (WORLD VIEW: A Personal Take on Events)," *Nature* 481, no. 7380 (2012): 115.

14. Ron A. M. Fouchier, Adolfo García-Sastre, and Yoshihiro Kawaoka, "Pause on Avian Flu Transmission Studies," *Nature* 481, no. 7382 (2012): 443, doi: 10.1038/481443a. The full list of authors is available in the online supplement to the article at https://static-content.springer.com/esm/art%3A10.1038%2F481443a/MediaObjects/41586_2012_BF481443a_MOESM219_ESM.pdf.

15. This and what follows is narrated in Jon Cohen, "WHO Group: H5N1 Papers Should Be Published in Full," *Science* 335, no. 6071 (2012): 899–900, doi:10.1126/science.335.6071.899.

16. Quoted in Jon Cohen and David Malakoff, "NSABB Members React to Request for Second Look at H5N1 Flu Studies—AAAS," *Science Insider*, March 2, 2012, https://www.science.org/content/article/nsabb-members-react-request-second-look-h5n1-flu-studies.

17. Francis Collins, "Statement by NIH Director Francis Collins, M.D., Ph.D. on the NSABB Review of Revised H5N1 Manuscripts," National Institutes of Health, March 30, 2015, https://www.nih.gov/about-nih/who-we-are/nih-director/statements/statement-nih-director-francis-collins-md-phd-nsabb-review-revised-h5n1-manuscripts.

18. See Nicholas G. Evans, Michael J. Selgelid, and Robert Mark Simpson, "Reconciling Regulation with Scientific Autonomy in Dual-Use Research," *Journal of Medicine and Philosophy: A Forum for Bioethics and Philosophy of Medicine* 47, no. 1 (February 1, 2022): 72–94, https://doi.org/10.1093/jmp/jhab041.

Chapter 5

1. Department of Health and Human Services, "A Framework for Guiding U.S. Department of Health and Human Services Funding Decisions about Research Proposals with the Potential for Generating Highly Pathogenic Avian Influenza H5N1 Viruses That Are Transmissible among Mammals by Respiratory Droplets," February 2013.

2. Harold Jaffe, Amy P. Patterson, and Nicole Lurie, "Extra Oversight for H7N9 Experiments," *Science* 341, no. 6147 (August 16, 2013): 713–714.

3. Department of Health and Human Services, "A Framework for Guiding U.S. Department of Health and Human Services Funding Decisions."

4. "High-Containment Biosafety Laboratories: DHS Lacks Evidence to Conclude That Foot-and-Mouth Disease Research Can Be Done Safely on the U.S. Mainland," GAO, May 22, 2008, https://www.gao.gov/products/gao-08-821t.

5. Marc Lipsitch and Barry R. Bloom, "Rethinking Biosafety in Research on Potential Pandemic Pathogens," *mBio* 3, no. 5 (2012), doi:10.1128/mBio.00360-12.

6. Lynn Klotz and Edward J. Sylvester, "The Unacceptable Risks of a Man-Made Pandemic," *Bulletin of the Atomic Scientists* (blog), August 7, 2012, https://thebulletin.org/2012/08/the-unacceptable-risks-of-a-man-made-pandemic/.

7. "CDC Lab Determines Possible Anthrax Exposures: Staff Provided Antibiotics/Monitoring," Centers for Disease Control and Prevention, June 19, 2014, https://www.cdc.gov/media/releases/2014/s0619-anthrax.html.

8. "Report on the Inadvertent Cross-Contamination and Shipment of a Laboratory Specimen with Influenza Virus H5N1," Centers for Disease Control and Prevention, August 2014, https://www.cdc.gov/about/pdf/lab-safety/investigationcdch5n1contaminationeventaugust15.pdf.

9. Joseph Tartal, *Corrective and Preventive Action Basics* (Silver Spring, MD: US Food and Drug Administration, 2014), https://www.fda.gov/files/about%20fda/published/CDRH-Learn-Presenation--Corrective-and-Preventive-Action-Basics.pdf.

10. Ian Sample, "Revealed: 100 Safety Breaches at UK Labs Handling Potentially Deadly Diseases," *Guardian*, December 4, 2014, https://www.theguardian.com/science/2014/dec/04/-sp-100-safety-breaches-uk-labs-potentially-deadly-diseases; Alison Young, *Pandora's Gamble: Lab Leaks, Pandemics, and a World at Risk* (Nashville: Center Street, 2023).

11. *Biosafety in Microbiological and Biomedical Laboratories (BMBL)*, 6th ed., Centers for Disease Control and Prevention, June 2020, https://www.cdc.gov/labs/BMBL.html, 11.

12. Rocco Cassagrande et al., *Assessing the Risks and Benefits of Conducting Research on Pathogens of Pandemic Potential* (Takoma Park, MD: Gryphon Scientific, 2015).

13. National Science Advisory Board for Biosecurity, "Recommendations for the Evaluation and Oversight of Proposed Gain-of-Function Research," May 2016, https://osp.od.nih.gov/wp-content/uploads/2016/06/NSABB_Final_Report_Recommendations_Evaluation_Oversight_Proposed_Gain_of_Function_Research.pdf, 43–44.

14. Office of Science and Technology Policy, "Recommended Policy Guidance for Departmental Development of Review Mechanisms for Potential Pandemic Pathogen Care and Oversight (P3CO)," January 9, 2017, 2–3.

15. Department of Health and Human Services, "Framework for Guiding Funding Decisions about Proposed Research Involving Enhanced Potential Pandemic Pathogens," 2017, https://www.phe.gov/s3/dualuse/Documents/P3CO.pdf, 3.

Chapter 6

1. NSABB, "Proposed Biosecurity Oversight Framework for the Future of Science," March 2023, https://osp.od.nih.gov/wp-content/uploads/2023/03/NSABB-Final-Report-Proposed-Biosecurity-Oversight-Framework-for-the-Future-of-Science.pdf.

2. Nicholas Greig Evans, "Great Expectations—Ethics, Avian Flu and the Value of Progress," *Journal of Medical Ethics* 39, no. 4 (March 2013): 210, https://doi.org/10.1136/medethics-2012-100712.

3. Nicholas G. Evans, *War on All Fronts: A Theory of Health Security Justice* (Cambridge, MA: MIT Press, 2023).

4. See, for example, Maggie Fox, "The World Is Unprepared for the Next Pandemic, Study Finds," CNN, December 9, 2021, https://www.cnn.com/2021/12/08/health/world-unprepared-pandemic-report/index.html; Andrew C. Heinrich and Saad B. Omer, "The World Isn't Ready for the Next Outbreak: The Case for a Pandemic Trust Fund," *Foreign Affairs*, September 6, 2021, https://www.foreignaffairs.com/articles/world/2021-09-06/world-isnt-ready-next-outbreak; Jennifer Kahn, "How Scientists Could Stop the Next Pandemic Before It Starts," *New York Times Magazine*, April 22, 2020, https://www.nytimes.com/2020/04/21/magazine/pandemic-vaccine.html; Bryan Walsh, "The World Is Not Ready for the Next Pandemic," *Time*, May 4, 2017, https://time.com/magazine/us/4766607/may-15th-2017-vol-189-no-18-u-s/; Myah Ward, "We're Not Ready for the Next Pandemic," *POLITICO*, September 24, 2021, https://politi.co/2Y1llB7. See also the excellent series on biodefense, biosecurity, and later health security spending trends assembled by the Johns Hopkins Center for Health Security, beginning with Ari Schuler, "Billions for Biodefense: Federal Agency Biodefense Funding, FY2001–FY2005," *Biosecurity and Bioterrorism: Biodefense Strategy, Practice, and Science* 2, no. 2 (June 2004): 86–96, https://doi.org/10.1089/153871304323146388.

5. See, for example, Langdon Winner, "Engineering Ethics and Political Imagination," in *Broad and Narrow Interpretations of Philosophy of Technology*, ed. Paul T. Durbin (Dordrecht: Springer, 1990), 53–64, https://doi.org/10.1007/978-94-009-0557-3_6.

6. Academic works to this effect include Christian Enemark, *Biosecurity Dilemmas: Dreaded Diseases, Ethical Responses, and the Health of Nations*, ill. ed. (Washington, DC: Georgetown University Press, 2017); Jeremy Youde, *Global Health Governance* (Cambridge, UK: Polity, 2012). In the United States, the Government Accountability Office has at times weighed in on these issues as well.

7. S. Schultz-Cherry et al., "Influenza Gain-of-Function Experiments: Their Role in Vaccine Virus Recommendation and Pandemic Preparedness," *mBio* 5, no. 6 (2014): doi:10.1128/mBio.02430-14.

8. Cambridge Working Group, "Cambridge Working Group Consensus Statement on the Creation of Potential Pandemic Pathogens (PPPs)," July 14, 2014, https://www.cambridgeworkinggroup.org/.

9. See Gryphon Scientific, *Risk and Benefit Analysis of Gain of Function Research* (Washington, DC: Gryphon Scientific, 2016), chapter 9.

10. Executive Office of the President of the United States, "United States Government Policy for Oversight of Dual Use Research of Concern and Pathogens

with Enhanced Pandemic Potential," May 6, 2024, https://www.whitehouse.gov/wp-content/uploads/2024/05/USG-Policy-for-Oversight-of-DURC-and-PEPP.pdf, 12.

11. Executive Office of the President of the United States, "United States Government Policy for Oversight of Dual Use Research of Concern and Pathogens with Enhanced Pandemic Potential," 13.

12. "NEIDL Researchers Refute UK Article about COVID Strain," *Brink*, October 17, 2022, https://www.bu.edu/articles/2022/neidl-researchers-refute-uk-article-about-covid-strain/.

13. Cathleen O'Grady, "'Overwhelmed by Hate': COVID-19 Scientists Face an Avalanche of Abuse, Survey Shows," *Science* 375, no. 6587 (March 25, 2022), https://www.science.org/content/article/overwhelmed-hate-covid-19-scientists-face-avalanche-abuse-survey-shows.

14. Department of Health and Human Services, "Framework for Guiding Funding Decisions about Proposed Research Involving Enhanced Potential Pandemic Pathogens," 2017, https://www.phe.gov/s3/dualuse/Documents/P3CO.pdf, 2.

15. Executive Office of the President of the United States, "United States Government Policy for Oversight of Dual Use Research of Concern and Pathogens with Enhanced Pandemic Potential," 9; emphasis in original.

16. Suzanne Goldenberg, "Bush Administration Accused of Doctoring Scientists' Reports on Climate Change," *Guardian*, January 31, 2007, https://www.theguardian.com/environment/2007/jan/31/usnews.frontpagenews.

17. Julie K. Pfeiffer, "Is the Debate and 'Pause' on Experiments That Alter Pathogens with Pandemic Potential Influencing Future Plans of Graduate Students and Postdoctoral Fellows?," *mBio* 6, no. 1 (2015): doi:10.1128/mBio.02525-14.

Chapter 7

1. Michael J. Selgelid, "Gain-of-Function Research: Ethical Analysis," *Science and Engineering Ethics* 22, no. 4 (August 2016): 923–964, https://doi.org/10.1007/s11948-016-9810-1.

2. Michelle Rozo and Gigi Kwik Gronvall, "The Reemergent 1977 H1N1 Strain and the Gain-of-Function Debate," ed. Mark R. Denison, *mBio* 6, no. 4 (September 2015): e01013–e01015, https://doi.org/10.1128/mBio.01013-15.

3. Gryphon Scientific, *Risk and Benefit Analysis of Gain of Function Research* (Washington, DC: Gryphon Scientific, 2016), §6.2.

4. Nicholas Greig Evans, "Great Expectations—Ethics, Avian Flu and the Value of Progress," *Journal of Medical Ethics* 39, no. 4 (March 2013): 209–213, https://doi.org/10.1136/medethics-2012-100712.

5. See Nicholas G. Evans, *War on All Fronts: A Theory of Health Security Justice* (Cambridge, MA: MIT Press, 2023), chapters 1–2.

6. Meena Krishnamurthy and Matthew Herder, "Justice in Global Pandemic Influenza Preparedness: An Analysis Based on the Values of Contribution, Ownership and Reciprocity," *Public Health Ethics* 6, no. 3 (November 1, 2013): 272–286, https://doi.org/10.1093/phe/pht027.

7. Lawrence O. Gostin et al., "Virus Sharing, Genetic Sequencing, and Global Health Security," *Science* 345, no. 6202 (September 12, 2014): 1295–1296, https://doi.org/10.1126/science.1257622; Alexandra L. Phelan et al., "Legal Agreements: Barriers and Enablers to Global Equitable COVID-19 Vaccine Access," *Lancet* 396, no. 10254 (September 19, 2020): 800–802, https://doi.org/10.1016/S0140-6736(20)31873-0.

8. Marc Lipsitch, Nicholas Greig Evans, and Owen Cotton Barratt, "Underprotection of Unpredictable Statistical Lives Compared to Predictable Ones," *Risk Analysis* 37, no. 5 (July 9, 2016): 893–904, https://doi.org/10.1111/risa.12658.

9. See, for example, Max Kozlov et al., "Biden Calls for Boosts in Science Spending to Keep US Competitive," *Nature* 615, no. 7953 (March 23, 2023): 572–573, https://doi.org/10.1038/d41586-023-00740-8; Joe Kennedy, "Strong Action Is Needed to Preserve the Competitiveness of America's Life-Science Industries," *STAT*, May 7, 2018, https://www.statnews.com/2018/05/07/life-sciences-america-preserve-competition/; Gigi Kwik Gronvall, "US Competitiveness in Synthetic Biology," *Health Security* 13, no. 6 (2015): 378–389, https://doi.org/10.1089/hs.2015.0046.

Chapter 8

1. Brett Edwards, *Insecurity and Emerging Biotechnology: Governing Misuse Potential* (Cham, Switz.: Palgrave, 2019).

2. Laura Spinney, *Pale Rider: The Spanish Flu of 1918 and How It Changed the World* (New York: PublicAffairs, 2017).

3. Nicholas G. Evans, Tara C. Smith, and Maimuna S. Majumder, *Ebola's Message: Public Health and Medicine in the Twenty-First Century* (Cambridge, MA: MIT Press, 2016).

4. Smriti Mallapaty, "COVID-Origins Study Links Raccoon Dogs to Wuhan Market: What Scientists Think," *Nature*, March 21, 2023, https://www.nature.com/articles/d41586-023-00827-2#:~:text=Raccoon%20dogs%2C%20bamboo%20rats%2C%20palm,2%2C%20which%20causes%20the%20disease.

5. Josh Rogin, "State Department Cables Warned of Safety Issues at Wuhan Lab Studying Bat Coronaviruses," *Washington Post*, April 14, 2020, https://

www.washingtonpost.com/opinions/2020/04/14/state-department-cables-warned-safety-issues-wuhan-lab-studying-bat-coronaviruses/.

6. everythingism (@_everythingism), "So first of all the market was identified as a pandemic risk by health officials half a decade before the SARS-CoV-2 outbreak. The virologist Eddie Holmes was taken there and he took pictures of raccoon dogs he saw being sold and slaughtered on site," X, August 30, 2022, 7:32 p.m., https://x.com/_everythingism/status/1564758024941871105.

7. Katie Rogers, Lara Jakes, and Ana Swanson, "Trump Defends Using 'Chinese Virus' Label, Ignoring Growing Criticism," *New York Times*, March 18, 2020, https://www.nytimes.com/2020/03/18/us/politics/china-virus.html.

8. Energy and Commerce Committee, "E&C Investigation Uncovers Earliest Known SARS-CoV-2 Sequence Released Outside of China," January 17, 2024, https://energycommerce.house.gov/posts/e-and-c-investigation-uncovers-earliest-known-sars-co-v-2-sequence-released-outside-of-china.

9. Cong Cao, "China's Evolving Biosafety/Biosecurity Legislations," *Journal of Law and the Biosciences* 8, no. 1 (April 10, 2021): lsab020, https://doi.org/10.1093/jlb/lsab020.

10. Gigi Kwik Gronvall et al., "The Biological Weapons Convention Should Endorse the Tianjin Biosecurity Guidelines for Codes of Conduct," *Trends in Microbiology* 30, no. 12 (December 2022): 1119–1120, https://doi.org/10.1016/j.tim.2022.09.014.

11. Stuart D. Blacksell et al., "Laboratory-Acquired Infections and Pathogen Escapes Worldwide between 2000 and 2021: A Scoping Review," *Lancet Microbe* 5, no. 2 (February 2024), https://doi.org/10.1016/S2666-5247(23)00319-1.

12. David P. Steensma and Robert A. Kyle, "Dr Li Wenliang: Wuhan 'Whistleblower' and Early COVID-19 Victim," *Mayo Clinic Proceedings* 97, no. 7 (July 2022): 1409–1410, https://doi.org/10.1016/j.mayocp.2022.05.033; Andrew Green, "Li Wenliang," *The Lancet* 395, no. 10225 (February 29, 2020): 682, https://doi.org/10.1016/S0140-6736(20)30382-2.

13. Jeremy Page Hinshaw, Betsy McKay, and Drew Hinshaw, "How the WHO's Hunt for Covid's Origins Stumbled in China," *Wall Street Journal*, March 17, 2021, World, https://www.wsj.com/articles/who-china-hunt-covid-origins-11616004512; Smriti Mallapaty, "Where Did COVID Come From? WHO Investigation Begins but Faces Challenges," *Nature*, November 11, 2020, https://doi.org/10.1038/d41586-020-03165-9.

14. Katherine Eban, "In Major Shift, NIH Admits Funding Risky Virus Research in Wuhan," *Vanity Fair*, October 22, 2021, https://www.vanityfair.com/news/2021/10/nih-admits-funding-risky-virus-research-in-wuhan.

15. Vineet D. Menachery et al., "A SARS-Like Cluster of Circulating Bat Coronaviruses Shows Potential for Human Emergence," *Nature Medicine* 21, no. 12 (December 2015): 1508–1513, https://doi.org/10.1038/nm.3985.

16. Da-Yuan Chen et al., "Role of Spike in the Pathogenic and Antigenic Behavior of SARS-CoV-2 BA.1 Omicron," bioRxiv (preprint), October 14, 2022, https://doi.org/10.1101/2022.10.13.512134.

17. Emery Winter, "No, Chinese Scientists Did Not 'Create COVID-19 Strain That Is 100% Lethal,'" News Center Maine, updated February 2, 2024, https://www.newscentermaine.com/article/news/verify/coronavirus-verify/chinese-scientists-didnt-create-mutant-covid-strain-100-percent-lethal-to-mice/536-e4f0d0eb-19a0-4911-b237-1a95a2d11442.

18. Tommy Tsan-Yuk Lam et al., "Identifying SARS-CoV-2-Related Coronaviruses in Malayan Pangolins," *Nature* 583, no. 7815 (July 2020): 282–285, https://doi.org/10.1038/s41586-020-2169-0.

19. Lai Wei et al., "An Infection and Pathogenesis Mouse Model of SARS-CoV-2-Related Pangolin Coronavirus GX_P2V(short_3UTR)," bioRxiv (preprint), January 21, 2024, https://doi.org/10.1101/2024.01.03.574008; Shanshan Lu et al., "Induction of Significant Neutralizing Antibodies against SARS-CoV-2 by a Highly Attenuated Pangolin Coronavirus Variant with a 104nt Deletion at the 3'-UTR," *Emerging Microbes and Infections* 12, no. 1 (December 2023): 2151383, https://doi.org/10.1080/22221751.2022.2151383.

20. Sheng Niu et al., "Molecular Basis of Cross-Species ACE2 Interactions with SARS-CoV-2-Like Viruses of Pangolin Origin," *EMBO Journal* 41, no. 1 (January 4, 2022): e109962, https://doi.org/10.15252/embj.2021109962; Mei-Qin Liu et al., "A SARS-CoV-2-Related Virus from Malayan Pangolin Causes Lung Infection without Severe Disease in Human ACE2-Transgenic Mice," *Journal of Virology* 97, no. 2 (February 28, 2023): e0171922, https://doi.org/10.1128/jvi.01719-22.

21. Nicholas G. Evans, *War on All Fronts: A Theory of Health Security Justice* (Cambridge, MA: MIT Press, 2023).

22. Vannevar Bush, *Science: The Endless Frontier*, 75th ann. ed. (Alexandria, VA: National Science Foundation, 2020), https://www.nsf.gov/about/history/EndlessFrontier_w.pdf, xiv, 17.

23. Bush, *Science*, xiii, 23–24.

FURTHER READING

Chapter 1

Arras, John D. "Getting Down to Cases: The Revival of Casuistry in Bioethics." *Journal of Medicine and Philosophy* 16, no. 1 (1991): 29–51. https://doi.org/10.1093/jmp/16.1.29.

Enserink, Martin. "Scientists Brace for Media Storm around Controversial Flu Studies." *Science Insider*, November 23, 2011. https://www.science.org/content/article/scientists-brace-media-storm-around-controversial-flu-studies.

Gillum, David, and Rebecca Moritz. "Why Gain-of-Function Research Matters." *Conversation*, June 21, 2021. http://theconversation.com/why-gain-of-function-research-matters-162493.

Hale, Benjamin G., John Steel, Rafael A. Medina, Balaji Manicassamy, Jianqiang Ye, Danielle Hickman, Rong Hai et al. "Inefficient Control of Host Gene Expression by the 2009 Pandemic H1N1 Influenza A Virus NS1 Protein." *Journal of Virology* 84, no. 14 (2010): 6909–6922. https://doi.org/10.1128/jvi.00081-10.

Herfst, Sander, Eefje J. A. Schrauwen, Martin Linster, Salin Chutinimitkul, Emmie de Wit, Vincent J. Munster, Erin M. Sorrell et al. "Airborne Transmission of Influenza A/H5N1 Virus between Ferrets." *Science* 336, no. 6088 (June 2012): 1534–1541. https://doi.org/10.1126/science.1213362.

Imai, Masaki, Tokiko Watanabe, Masato Hatta, Subash C. Das, Makoto Ozawa, Kyoko Shinya, Gongxun Zhong et al. "Experimental Adaptation of an Influenza H5 HA Confers Respiratory Droplet Transmission to a Reassortant H5 HA/H1N1 Virus in Ferrets." *Nature* 486, no. 7403 (2012): 420–428. https://doi.org/10.1038/nature10831.

National Research Council. *Potential Risks and Benefits of Gain-of-Function Research: Summary of a Workshop*. Washington, DC: National Academies Press, 2015.

Saalbach, K. P. "Gain-of-Function Research." In *Advances in Applied Microbiology*, edited by Geoffrey Michael Gadd and Sima Sariaslani, 120:79–111. Cambridge, MA: Academic Press, 2022. https://doi.org/10.1016/bs.aambs.2022.06.002.

Chapter 2

"Amerithrax Investigative Summary—United States Department of Justice." US Department of Justice Archives, October 15, 2010. https://www.justice.gov/archive/amerithrax/docs/amx-investigative-summary.pdf.

Cello, Jeronimo, Aniko V. Paul, and Eckard Wimmer. "Chemical Synthesis of Poliovirus CDNA: Generation of Infectious Virus in the Absence of Natural Template." *Science* 297, no. 5583 (2002): 1016–1018.

Enemark, Christian. "United States Biodefense, International Law, and the Problem of Intent." *Politics and the Life Sciences* 24, no. 1–2 (2005): 32–42.

Gillispie, Charles Coulston. "Science and Secret Weapons Development in Revolutionary France, 1792–1804: A Documentary History." *Historical Studies in the Physical and Biological Sciences* 23, no. 1 (1992): 35–152. https://doi.org/10.2307/27757692.

Jackson, Ronald J., Deborah J. Maguire, Lyn A. Hinds, and Ian A. Ramshaw. "Infertility in Mice Induced by a Recombinant Ectromelia Virus Expressing Mouse Zona Pellucida Glycoprotein 3." *Biology of Reproduction* 58, no. 1 (January 1998): 152–159. doi:10.1095/biolreprod58.1.152.

National Research Council. *Biotechnology Research in an Age of Terrorism*. Washington, DC: National Academies Press, 2004. https://doi.org/10.17226/10827.

National Research Council. *Scientific Communication and National Security*. Washington, DC: National Academies Press, 1982. https://doi.org/10.17226/253.

Pollack, Andrew. "Scientists Create a Live Polio Virus." *New York Times*, July 2, 2002.

Tartal, Joseph. *Corrective and Preventive Action Basics*. Silver Spring, MD: US Food and Drug Administration, 2014. https://www.fda.gov/files/about%20fda/published/CDRH-Learn-Presenation--Corrective-and-Preventive-Action-Basics.pdf.

Taubenberger, Jeffery K., Ann H. Reid, Raina M. Lourens, Ruixue Wang, Guozhong Jin, and Thomas G. Fanning. "Characterization of the 1918 Influenza Virus Polymerase Genes." *Nature* 437, no. 7060 (2005): 889–893. doi:10.1038/nature04230.

Tumpey, Terrence M., Christopher F. Basler, Patricia V. Aguilar, Hui Zheng, Alicia Solórzano, David E. Swayne, Nancy J. Cox et al. "Characterization of the

Reconstructed 1918 Spanish Influenza Pandemic Virus." *Science* 310 (2005): 77–80. doi:10.1126/science.1119392.

Wright, Susan. "Taking Biodefense Too Far." *Bulletin of the Atomic Scientists* 60, no. 6 (2004): 58–66. doi:10.2968/060006013.

Chapter 3

Anderson, Tavis K., Catherine A. Macken, Nicola S. Lewis, Richard H. Scheuermann, Kristien Van Reeth, Ian H. Brown, Sabrina L. Swenson et al. "A Phylogeny-Based Global Nomenclature System and Automated Annotation Tool for H1 Hemagglutinin Genes from Swine Influenza A Viruses." *mSphere* 1, no. 6 (2016): e00275–16. https://doi.org/10.1128/mSphere.00275-16.

Belser, Jessica A., Alissa M. Eckert, Thanhthao Huynh, Joy M. Gary, Jana M. Ritter, Terrence M. Tumpey, and Taronna R. Maines. "A Guide for the Use of the Ferret Model for Influenza Virus Infection." *American Journal of Pathology* 190, no. 1 (2020): 11–24. doi:10.1016/j.ajpath.2019.09.017.

Dietrich, Michael R., Rachel A. Ankeny, Nathan Crowe, Sara Green, and Sabina Leonelli. "How to Choose Your Research Organism." *Studies in History and Philosophy of Biological and Biomedical Sciences* 80 (April 2020): 101227. doi:10.1016/j.shpsc.2019.101227.

Herfst, Sander, Eefje J. A. Schrauwen, Martin Linster, Salin Chutinimitkul, Emmie de Wit, Vincent J. Munster, Erin M. Sorrell et al. "Airborne Transmission of Influenza A/H5N1 Virus between Ferrets." *Science* 336, no. 6088 (2012): 1534–1541. doi:10.1126/science.1213362.

Imai, Masaki, Tokiko Watanabe, Masato Hatta, Subash C. Das, Makoto Ozawa, Kyoko Shinya, Gongxun Zhong et al. "Experimental Adaptation of an Influenza H5 HA Confers Respiratory Droplet Transmission to a Reassortant H5 HA /H1N1 Virus in Ferrets." *Nature* 486, no. 7403 (2012): 420–428. doi:10.1038/nature10831.

Oh, Ding Y., and Aeron C. Hurt. "Using the Ferret as an Animal Model for Investigating Influenza Antiviral Effectiveness." *Frontiers in Microbiology* 7 (February 2016): 80. doi:10.3389/fmicb.2016.00080.

"A Revision of the System of Nomenclature for Influenza Viruses: A WHO Memorandum." *Bulletin of the World Health Organization* 58, no. 4 (1980): 585–591.

Scheurmann, Richard H. "Global H1 Influenza Nomenclature." J. Craig Venter Institute, 2016. https://www.jcvi.org/research/global-h1-influenza-nomenclature.

Smith, Wilson, C. H. Andrewes, and P. P. Laidlaw. "A Virus Obtained from Influenza Patients." *Lancet* 222, no. 5732 (1933): 66–68. doi:10.1016/S0140-6736(00)78541-2.

Chapter 4

Berns, Kenneth I., Arturo Casadevall, Murray L. Cohen, Susan A. Ehrlich, Lynn W. Enquist, J. Patrick Fitch, David R. Franz et al. "Adaptations of Avian Flu Virus Are a Cause for Concern." *Science* 335, no. 6069 (2012): 660–661. doi:10.1126/science.1217994.

Cohen, Jon. "WHO Group: H5N1 Papers Should Be Published in Full." *Science* 335, no. 6071 (2012): 899–900. doi:10.1126/science.335.6071.899.

Cohen, Jon, and David Malakoff. "NSABB Members React to Request for Second Look at H5N1 Flu Studies—AAAS." *Science Insider*, March 2, 2012. https://www.science.org/content/article/nsabb-members-react-request-second-look-h5n1-flu-studies.

Fouchier, Ron A. M., Adolfo García-Sastre, and Yoshihiro Kawaoka. "Pause on Avian Flu Transmission Studies." *Nature* 481, no. 7382 (2012): 443. doi:10.1038/481443a.

Harmon, Katherine. "What Really Happened in Malta This September When Contagious Bird Flu Was First Announced?" *Scientific American*, December 30, 2011. https://www.nih.gov/about-nih/who-we-are/nih-director/statements/statement-nih-director-francis-collins-md-phd-nsabb-review-revised-h5n1-manuscripts.

Kennedy, Donald. "Editorial: Better Never than Late." *Science* 310, no. 5746 (2005): 195.

National Research Council. *Perspectives on Research with H5N1 Avian Influenza: Scientific Inquiry, Communication, Controversy: Summary of a Workshop.* Washington, DC: National Academies Press, 2013.

Palese, Peter. "Don't Censor Life-Saving Science: Controlling Who Is Allowed Access to Information about Mutations in the H5N1 Bird Flu Virus Is Unacceptable (WORLD VIEW: A Personal Take on Events)." *Nature* 481, no. 7380 (2012): 115.

"Press Statement on the NSABB Review of H5N1 Research." National Institutes of Health, September 18, 2015. https://www.nih.gov/news-events/news-releases/press-statement-nsabb-review-h5n1-research.

Shattuck, Roger. *Forbidden Knowledge: From Prometheus to Pornography*. San Diego: Harcourt Brace and Co., 1997.

Chapter 5

Biosafety in Microbiological and Biomedical Laboratories (BMBL). 6th ed. Centers for Disease Control and Prevention, June 2020, https://www.cdc.gov/labs/BMBL.html.

Cassagrande, Rocco, Casey Basham, Emily Billings, Audrey Cereles, Mikaela Finnegan, Mark Kazmierczak, Erin Lauer et al. *Assessing the Risks and Benefits of Conducting Research on Pathogens of Pandemic Potential*. Takoma Park, MD: Gryphon Scientific, 2015.

"CDC Lab Determines Possible Anthrax Exposures: Staff Provided Antibiotics/Monitoring." Centers for Disease Control and Prevention, June 19, 2014. https://www.cdc.gov/media/releases/2014/s0619-anthrax.html.

"Deliberative Process and Funding Pause on Certain Types of Gain-of-Function Research." Science Safety Security, October 16, 2014. https://www.phe.gov/s3/dualuse/Pages/Deliberative-Process-GOF.aspx.

"Framework for Guiding Funding Decisions about Proposed Research Involving Enhanced Potential Pandemic Pathogens." US Department of Health and Human Services, 2017. https://www.phe.gov/s3/dualuse/Documents/p3co.pdf.

Lipsitch, Marc, and Barry R. Bloom. "Rethinking Biosafety in Research on Potential Pandemic Pathogens." *mBio* 3, no. 5 (2012). doi:10.1128/mBio.00360-12.

Patterson, Amy P., Lawrence A. Tabak, Anthony S. Fauci, Francis S. Collins, and Sally Howard. "A Framework for Decisions about Research with HPAI H5N1 Viruses." *Science* 339, no. 6123 (2013): 1036–1037. doi:10.1126/science.1236194.

"Recommended Policy Guidance for Departmental Development of Review Mechanisms for Potential Pandemic Pathogen Care and Oversight (P3CO)." January 9, 2017. https://www.phe.gov/s3/dualuse/Documents/P3CO-Final GuidanceStatement.pdf.

Sample, Ian. "Revealed: 100 Safety Breaches at UK Labs Handling Potentially Deadly Diseases." *Guardian*, December 4, 2014. https://www.theguardian.com/science/2014/dec/04/-sp-100-safety-breaches-uk-labs-potentially-deadly-diseases.

Selgelid, Michael J. "Gain-of-Function Research: Ethical Analysis." *Science and Engineering Ethics* 22, no. 4 (2016): 923–964. doi:10.1007/s11948-016-9810-1.

Chapter 6

Branswell, Helen. "Boston University Researchers' Testing of Lab-Made Version of Covid Virus Draws Government Scrutiny." *STAT* (blog), October 18, 2022. https://www.statnews.com/2022/10/17/boston-university-researchers-testing-of-lab-made-version-of-covid-virus-draws-government-scrutiny/.

"Cambridge Working Group Consensus Statement on the Creation of Potential Pandemic Pathogens (PPPs)." July 14, 2014. http://www.cambridgeworkinggroup.org/documents/statement.pdf.

Pfeiffer, Julie K. "Is the Debate and 'Pause' on Experiments That Alter Pathogens with Pandemic Potential Influencing Future Plans of Graduate Students and Postdoctoral Fellows?" *mBio* 6, no. 1 (2015). doi:10.1128/mBio.02525-14.

Schultz-Cherry, S., R. J. Webby, R. G. Webster, A. Kelso, I. G. Barr, J. W. McCauley, R. S. Daniels et al. "Influenza Gain-of-Function Experiments: Their Role in Vaccine Virus Recommendation and Pandemic Preparedness." *mBio* 5, no. 6 (2014). doi:10.1128/mBio.02430-14.

Wang, Weijia, Bin Lu, Helen Zhou, Amorsolo L. Suguitan Jr., Xing Cheng, Kanta Subbarao, George Kemble, and Hong Jin. "Glycosylation at 158N of the Hemagglutinin Protein and Receptor Binding Specificity Synergistically Affect the Antigenicity and Immunogenicity of a Live Attenuated H5N1 A/Vietnam/1203/2004 Vaccine Virus in Ferrets." *Journal of Virology* 84, no. 13 (2010): 6570–6577. doi:10.1128/JVI.00221-10.

Yamada, Shinya, Yasuo Suzuki, Takashi Suzuki, Mai Q. Le, Chairul A. Nidom, Yuko Sakai-Tagawa, Yukiko Muramoto et al. "Haemagglutinin Mutations Responsible for the Binding of H5N1 Influenza A Viruses to Human-Type Receptors." *Nature* 444, no. 7117 (2006): 378–382. doi:10.1038/nature05264.

Yang, Zhi-Yong, Chih-Jen Wei, Wing-Pui Kong, Lan Wu, Ling Xu, David F. Smith, and Gary J. Nabel. "Immunization by Avian H5 Influenza Hemagglutinin

Mutants with Altered Receptor Binding Specificity." *Science* 317, no. 5839 (2007): 825–828. doi:10.1126/science.1135165.

Chapter 7

Evans, Nicholas Greig. "Ethical and Philosophical Considerations for Gain-of-Function Policy: The Importance of Alternate Experiments." *Frontiers in Bioengineering and Biotechnology* 6 (February 2018): e1875. https://doi.org/10.3389/fbioe.2018.00011.

Evans, Nicholas Greig. "Great Expectations—Ethics, Avian Flu and the Value of Progress." *Journal of Medical Ethics* 39, no. 4 (March 2013): 209–213. https://doi.org/10.1136/medethics-2012-100712.

Evans, Nicholas Greig, Marc Lipsitch, and Meira Levinson. "The Ethics of Biosafety Considerations in Gain-of-Function Research Resulting in the Creation of Potential Pandemic Pathogens." *Journal of Medical Ethics* 41, no. 11 (November 2015): 901–908. https://doi.org/10.1136/medethics-2014-102619.

Krishnamurthy, Meena, and Matthew Herder. "Justice in Global Pandemic Influenza Preparedness: An Analysis Based on the Values of Contribution, Ownership and Reciprocity." *Public Health Ethics* 6, no. 3 (November 1, 2013): 272–286. https://doi.org/10.1093/phe/pht027.

Lipsitch, Marc, Nicholas Greig Evans, and Owen Cotton Barratt. "Underprotection of Unpredictable Statistical Lives Compared to Predictable Ones." *Risk Analysis* 37, no. 5 (July 9, 2016): 893–904. https://doi.org/10.1111/risa.12658.

Selgelid, Michael J. "Gain-of-Function Research: Ethical Analysis." *Science and Engineering Ethics* 22, no. 4 (August 2016): 923–964. https://doi.org/10.1007/s11948-016-9810-1.

INDEX

Asilomar Conference, 24, 82

Biosafety, 92, 102, 129, 165, 169–170
Biosecurity, 38, 120, 129, 169

Censorship, 39, 79, 83, 148
Centers for Disease Control and Prevention, 30, 39–40, 101–102
COVID-19, alleged lab origin of, 165–170
Criteria for review of GOF studies
 first H5N1 policy, 97–98
 NSABB's proposed revision, 106–108
 P3CO policy, 108–110
 2024 policy, 129–131

Dual-use research
 definition of, 23–24
 Fink Report and, 34–37
 2024 policy on, 129–132

Ethics
 animal, 62, 92
 Belmont Report, 116
 criteria for review and, 110, 112
 definition of, 143–144
 GOF white paper, 105, 108
 and public good of science, 150–154
Experiments of concern, 35–37

Fauci, Anthony, 2, 81, 87, 95, 96
Fouchier, Ron, 6–8, 24, 48, 68, 83
 at ESWI working group, 78

Hemagglutinin, 10, 47, 53, 126

Influenza
 H5N1's fatality rate, 8
 naming conventions, 47
 role of the ferret in research, 59–64

Kawaoka, Yoshihiro, 13–14, 54–59, 86–87, 121
Keim, Paul, 79, 87

Neuraminidase, 10, 47, 126
NSABB
 changing position on H5N1 papers, 87–88
 creation of, 37
 H1N1 and, 40–41
 recommending against publication of H5N1 papers, 80–82
 role in the deliberative process, 105–108
Nuclear energy, 24–25

Palese, Peter, 83
Paul, Rand, 1, 182–183
Policy terms, 19–20, 91, 126, 179

Political imagination, 120–121
Proliferation of labs, 102

"Reasonably anticipated," as criteria for review, 132, 136–137

Transparency, 117, 119, 131–133
2024 dual-use policy, 129–132

World Health Organization (WHO/OMS), 38, 80, 85–88, 124

NICHOLAS G. EVANS is Associate Professor and Chair of Philosophy at the University of Massachusetts Lowell, and affiliate faculty at the Center for Terrorism and Security Studies. A 2020–2023 Greenwall Foundation Faculty Scholar, he is best known for his research on health security, emerging technologies, and performance enhancement in the military. His 2016 collection, *Ebola's Message: Public Health and Medicine in the Twenty-First Century*, received favorable reviews in *Nature*. His new, solely authored work on the ethics of pandemic preparedness and response, *War on All Fronts: A Theory of Health Security Justice*, was published with the MIT Press on May 16, 2023. Evans resides in Lowell with his wife and three cats: Lexi, Inky, and Joseph.